Math and Test Taking Grade 4
A Best Value Book™

Written by
Patricia Pedigo and Roger DeSanti

Edited by
Kelley Wingate Levy

Copyright © 2000, Carson-Dellosa Publishing Company, Inc., Greensboro, North Carolina 27425, publishers of the "Stick Out Your Neck" series. All rights reserved. The purchase of this material entitles the buyer to reproduce worksheets and activities for classroom use only—not for commercial resale. Reproduction of these materials for an entire school or district is prohibited. No part of this book may be reproduced (except as noted above), stored in a retrieval system, or transmitted in any form or by any means (mechanically, electronically, recording, etc.) without the prior written consent of Carson-Dellosa Publishing Co., Inc.

© Carson-Dellosa CD-3754

ISBN 0-88724-535-8

Table of Contents

Number Concepts
Number Words..1
Number Words..2
Sequence..3
Sequence..4
Sequence..5
Place Value 6
Place Value 7
Place Value 8
Place Value 9
Place Value 10
Fractions 11
Fractions 12
Fractions 13
Operational Signs 14
Operational Signs 15
Greater Than, Less Than, Equal To..........16
Mathematical Expressions.........................17
Mathematical Expressions.........................18
Mathematical Expressions.........................19
Mathematical Expressions.........................20
Mathematical Expressions.........................21
Mathematical Expressions.........................22
Fact Families 23
Fact Families 24
Number Concepts Practice 25-30

Computation
Two-Digit Addition 31
Two-Digit Addition 32
Two- and Three-Digit Addition 33
Two- and Three-Digit Addition 34
Three- and Four-Digit Addition 35
Three- and Four-Digit Addition 36
Addition with Three Addends....................37
Addition with Three Addends....................38
Two-Digit Subtraction..................................39
Two-Digit Subtraction..................................40
Two- and Three-Digit Subtraction 41
Two- and Three-Digit Subtraction 42
Three- and Four-Digit Subtraction...........43
Three- and Four-Digit Subtraction...........44
Multiplication (Facts to 12).........................45
Multiplication (Facts to 12).........................46

Multiplication (Facts to 12).........................47
Multiplication (Two-Digit Numbers) 48
Multiplication (Two-Digit Numbers) 49
Multiplication (Two Digit Numbers) 50
Division (Facts to 12)...................................51
Division (Facts to 12)...................................52
Division..53
Division..54
Missing Numbers (Multiplication) 55
Missing Numbers(Division) 56
Rounding Numbers 57
Rounding Numbers 58
Decimals/Fractions 59
Decimals/Fractions 60
Computation Practice..........................61-66

Application
Word Problems 67
Word Problems 68
Word Problems 69
Word Problems...70
Word Problems 71
Shapes..72
Perimeter 73
Area 74
Time...75
Time...76
Money..77
Money..78
Metric Conversion 79
Metric Measurement 80
Metric Measurement 81
Bar Graph 82
Pie Graph...83
Pictograph 84
Pictograph 85
Application Practice 86-91

Practice Tests
Practice Test......................................92-115
Answer Key......................................116-121
Student Answer Sheet 122-123

© Carson-Dellosa CD-3754

Math and Test Taking

Written by Patricia Pedigo and Roger DeSanti
Edited by Kelley Wingate Levy

About the Authors

Patricia Pedigo has many years of teaching experience in urban, rural, public, and private settings. She has taught at all elementary and middle school grade levels, and worked as a reading specialist for learning different students. Patricia has created reading materials that integrate content areas with language development skills. She holds an M.Ed. in Reading Education and is nearing completion of her doctoral studies.

Dr. Roger DeSanti has been an educator since the mid 1970s. His teaching experiences span a wide range of grades and ability levels, from deaf nurseries through university graduate schools. As a professor, he has authored numerous articles and books, achievement tests, and instructional materials.

Perfect for school or home, every **Kelley Wingate Best Value Book**™ has been designed to help students master the skills necessary to succeed. Each book is packed with reproducible test pages, 96 cut-apart flash cards, and supplemental resource pages full of valuable information, ideas, and activities. These activities may be used as classroom or homework activities, or as enrichment material for a math program.

The purpose of this book is to provide conceptual, computational, and applied mathematical skills practice while reinforcing positive test taking strategies. The format and types of activities have been patterned after those in standardized tests such as Stanford Achievement, LEAP, Iowa Test of Basic Skills, and other state and nationally based achievement tests. The activities have been sequenced to facilitate successful completion of the assigned tasks, thus building the confidence and self-esteem students need to meet academic challenges.

The practice tests in this book cover the range of cognitive skills from basic concepts, to computation skills, to applied mathematical concepts. Practice tests at the end of each section provide opportunities for cumulative review. These practice tests may be administered in various ways. One method is to give students the activities consecutively, each cognitive skill building upon the previous. Alternately, select one sheet from each skill category and give students this packet, a comprehensive approach similar to many standardized tests. Once students are familiar with the practice test format, consider giving them timed practice tests, since many standardized tests are timed. Take into consideration the length and difficulty of the test, as well as the competency level of the test-takers. Cumulative practice tests are included at the end of each skill section, and at the end of the book. Extra tests may be assembled by selecting one or two pages from each of the three skill areas and administering them together.

Test-Taking Strategies

Below are some suggestions for improving students' test-taking performance; you may wish to share some or all of these with your students:

- Keep a few pieces of scratch paper beside your test for working out problems.

- Use a ruler or index card to keep your place on the answer sheet.

- Have an extra sharpened pencil and a good eraser ready.

- Read the problems carefully. Decide if you know the answer before you look at the choices.

- Read all the choices, even if the first seems correct. If you don't know which answer is right, cross off the ones you know are wrong. Then, pick from the remaining choices.

- If your answer is not one of the choices, go back over your work. If you are certain your answer is correct, mark the space for NH (not here).

- Answers should be marked by completely darkening the bubble that corresponds to the selected choice.

- Always put down an answer. (If you aren't sure about the answer, make a good guess! If you leave a problem blank, you know it will be wrong. A guess might just be correct!)

Using the Answer Sheet

Included at the end of this book, preceding the flash cards, are universal answer sheets designed specifically for use with this book. Each sheet has enough answer columns for multiple test pages. When administering a test, reproduce this page for your students, making sure to specify at the top of each answer column the page number corresponding to the particular test they are taking. Optionally, these answer sheets can be reused until every column is filled. Space is also provided for test-scoring feedback.

Flash Card Ideas and Activities

Included in the back of this book are 96 flash cards ideal for individual review, group solving sessions, or as part of timed, sequential, or grouped tests. Pull out the flash cards and cut them apart or, if you have access to a paper cutter, use that to cut them into individual cards. Here are just a few of the ways you may want to use these flash cards:

- Play "Around the World with Flash Cards," the object of which is to be the first student to circle the room and return to his own seat. Have two students stand at their desks. Show them one flash card. The first to correctly answer the flash card problem advances and stands beside the next seat in line. The student in that seat stands. Show these two students a flash card, and repeat the process. The winner continues to advance and challenge other students until he returns to his own desk or until he loses the challenge, in which case he sits down and the new winner advances to face a new flash card and the next student.

- Hold a "Math Challenge." Divide the class into two teams and have a representative from each team stand at the front of the room beside a desk and attempt to be the first to answer a flash card problem. The student who is able to answer the question first can tap the desk (or ring a bell) to signal readiness to give an answer. Award points for correct answers. If the student answers incorrectly, allow the other team a chance to answer.

- Use a timer or stopwatch to record how many problems a student can answer correctly in a certain amount of time. Review incorrect answers and repeat the exercise. Provide rewards for improved scores.

- Give students a card with a math fact on it. Have them write out the other members of that fact family, or brainstorm other math problems that have the same answer.

- Use flash cards as impromptu quizzes. Give each student three to five cards attached to an answer sheet that he can complete and return. Vary the selection of cards given to each student for each quiz.

- Post a certain number of cards, daily or weekly, as bonus questions or for extra credit.

Name _____   Skill: Number Words

Directions
Read each question and choose the correct answer. Mark the space for the answer you have chosen. Mark NH if the answer is not here.

1. What is the numeral for eight thousand, one hundred eighteen?

   a. 801,081
   b. 80,118
   c. 8,181
   d. (8,118)
   e. NH

2. What is the numeral for sixteen thousand, nineteen?

   f. 1,619
   g. 16,109
   h. (16,019)
   j. 16,190
   k. NH

3. What is the numeral for four thousand, eighty-eight?

   a. 4,808
   b. 40,808
   c. 4,880
   d. (4,088)
   e. NH

4. What is the numeral for seventy thousand, one hundred forty-three?

   f. 71,043
   g. (70,143)
   h. 71,403
   j. 700,143
   k. NH

5. What is the numeral for sixty-four thousand, five hundred three?

   a. 6,453
   b. (64,503)
   c. 60,453
   d. 64,530
   e. NH

6. What is the numeral for twenty-one thousand, two hundred fifty-six?

   f. 21,526
   g. (21,256)
   h. 21,652
   j. 31,265
   k. NH

7. What is the numeral for forty-five thousand, five hundred twenty?

   a. 45,052
   b. 40,552
   c. 45,502
   d. (45,520)
   e. NH

8. What is the numeral for two hundred sixty thousand, nine hundred one?

   f. 206,901
   g. 260,910
   h. (260,901)
   j. 260,091
   k. NH

© Carson-Dellosa CD-3754

Name _____   Skill: Number Words

Directions
Read each question and choose the correct answer. Mark the space for the answer you have chosen. Mark NH if the answer is not here.

1. What is the numeral for two hundred seventy thousand, eight hundred thirty?

   a. (270,830)
   b. 207,830
   c. 270,803
   d. 27,830
   e. NH

2. What is the numeral for seventy-one thousand, thirteen?

   f. 7,113
   g. 71,103
   h. (71,013)
   j. 71,310
   k. NH

3. What is the numeral for forty-four thousand, three hundred twenty?

   a. 44,302
   b. 44,032
   c. (44,320)
   d. 40,432
   e. NH

4. What is the numeral for nine hundred thirty-three thousand?

   f. 900,330
   g. (933,000)
   h. 930,300
   j. 903,300
   k. NH

5. What is the numeral for four hundred ten thousand, seven hundred six?

   a. 401,760
   b. 410,076
   c. 410,760
   d. (410,706)
   e. NH

6. What is the numeral for four million, two hundred three thousand, seven hundred sixty-eight?

   f. 4,230,768
   g. (4,203,768)
   h. (4,023,768)
   j. 40,230,768
   k. NH

7. What is the numeral for twenty-five million, seventeen thousand, one hundred fifty-seven?

   a. 25,107,157
   b. 25,170,157
   c. (25,017,157)
   d. 2,517,157
   e. NH

8. What is the numeral for eleven million, five hundred thirty thousand, one hundred twelve?

   f. (11,530,112)
   g. 11,503,112
   h. 11,053,112
   j. 1,153,112
   k. NH

© Carson-Dellosa CD-3754

Name _____   Skill: Sequence

Directions
Read each question and choose the correct answer. Mark the space for the answer you have chosen. Mark NH if the answer is not here.

1. What is the missing number?

| 90 | 88 | 86 | 84 | 82 |

a. 92
b. 83
c. 82
d. 80
e. NH

5. What is the missing number?

| 620 | 520 | 420 | 320 | 220 |

a. 220
b. 300
c. 210
d. 319
e. NH

2. What is the missing number?

| 185 | 190 | 195 | 200 | 205 |

f. 191
g. 193
h. 199
j. 195
k. NH

6. What is the missing number?

| 1,000 | 995 | 990 | 985 | 980 |

f. 999
g. 1,001
h. 1,010
j. 995
k. NH

3. What is the missing number?

| 700 | 710 | 720 | 730 | 740 |

a. 731
b. 740
c. 750
d. 800
e. NH

7. What is the missing number?

| 425 | 450 | 475 | 500 | 525 |

a. 524
b. 500
c. 495
d. 476
e. NH

4. What is the missing number?

| 36 | 39 | 42 | 45 | 48 |

f. 42
g. 44
h. 41
j. 43
k. NH

8. What is the missing number?

| 983 | 883 | 783 | 683 | 583 |

f. 683
g. 584
h. 773
j. 680
k. NH

© Carson-Dellosa CD-3754

Name _____   Skill: Sequence

Directions
Read each question and choose the correct answer. Mark the space for the answer you have chosen. Mark NH if the answer is not here.

1. What number is greater than 1,483?

a. (1,843)
b. 1,348
c. 1,438
d. 1,393
e. NH

2. What number is greater than 6,707?

f. (6,770)
g. 6,077
h. 6,007
j. 6,700
k. NH

3. What number is greater than 12,465?

a. (14,256)
b. 12,300
c. 12,456
d. 11,999
e. NH

4. What number is greater than 156,321?

f. 156,231
g. 153,621
h. 156,299
j. (165,123)
k. NH

5. What number is less than 3,001?

a. 3,100
b. 3,101
c. (3,000)
d. 3,011
e. NH

6. What number is less than 5,780?

f. 7,580
g. (5,708)
h. 5,807
j. 5,791
k. NH

7. What number is less than 23,086?

a. 23,608
b. 32,068
c. 23,806
d. (23,068)
e. NH

8. What number is less than 105,293?

f. 501,000
g. 152,093
h. (105,239)
j. 105,393
k. NH

© Carson-Dellosa CD-3754

Name _____   Skill: Sequence

Directions
Read each question and choose the correct answer. Mark the space for the answer you have chosen. Mark NH if the answer is not here.

1. What number comes between 800 and 1,540?

   a. 774
   b. 1,290
   c. 1,590
   d. 1,600
   e. NH

2. What number comes between 659 and 723?

   f. 587
   g. 601
   h. 695
   j. 732
   k. NH

3. What number comes between 2,890 and 3,908?

   a. 1,980
   b. 2,809
   c. 2,969
   d. 3,980
   e. NH

4. What number comes between 4,876 and 5,011?

   f. 4,786
   g. 4,998
   h. 5,101
   j. 5,110
   k. NH

5. What number comes between 15,005 and 15,505?

   a. 15,004
   b. 15,050
   c. 15,550
   d. 15,515
   e. NH

6. What number comes between 17,855 and 19,677?

   f. 18,016
   g. 16,901
   h. 19,776
   j. 17,588
   k. NH

7. What number comes between 120,000 and 147,000?

   a. 102,000
   b. 210,000
   c. 174,000
   d. 142,000
   e. NH

8. What number comes between 929,900 and 1,000,000?

   f. 690,599
   g. 906,221
   h. 948,983
   j. 1,000,001
   k. NH

© Carson-Dellosa CD-3754

Name _____   Skill: Place Value

Directions
Read each question and choose the correct answer. Mark the space for the answer you have chosen. Mark NH if the answer is not here.

1. What means the same as 72?	5. What means the same as 759?
a. 5 tens and 12 ones b. 7 tens and 12 ones c. 2 tens and 7 ones d. 6 tens and 12 ones e. NH	a. 7 hundreds, 5 tens, and 19 ones b. 7 hundreds, 6 tens, and 9 ones c. 7 hundreds, 4 tens, and 19 ones d. 6 hundreds, 5 tens, and 4 ones e. NH
2. What means the same as 83?	6. What means the same as 912?
f. 3 tens and 8 ones g. 7 tens and 13 ones h. 6 tens and 13 ones j. 8 tens and 13 ones k. NH	f. 8 hundreds, 1 ten, and 2 ones g. 9 hundreds, 0 tens, and 12 ones h. 9 hundreds, 1 ten, and 0 ones j. 9 hundreds, 2 tens, and 1 ones k. NH
3. What means the same as 308?	7. What means the same as 397?
a. 2 tens and 8 ones b. 3 tens and 8 ones c. 3 hundreds, 8 tens, and 0 ones d. 3 hundreds, 0 tens, and 8 ones e. NH	a. 9 hundreds, 3 tens, and 7 ones b. 3 hundreds, 8 tens, and 17 ones c. 3 hundreds, 7 tens, and 17 ones d. 3 hundreds, 9 tens, and 17 ones e. NH
4. What means the same as 548?	8. What means the same as 609?
f. 5 hundreds, 3 tens, and 18 ones g. 5 hundreds, 2 tens, and 18 ones h. 5 hundreds, 4 tens, and 18 ones j. 5 hundreds, 5 tens, and 8 ones k. NH	f. 6 hundreds, 0 tens, and 0 ones g. 6 tens and 9 ones h. 6 hundreds, 9 tens, and 1_ j. 9 hundreds, 8 tens, an_ k. NH

© Carson-Dellosa CD-3754

Name _____   Skill: Place Value

Directions
Read each question and choose the correct answer. Mark the space for the answer you have chosen. Mark NH if the answer is not here.

---

1. What means the same as 74?

a. 700 + 4
b. 7 + 4
c. 70 + 4
d. 700 + 40
e. NH

2. What means the same as 380?

f. 300 + 80
g. 30 + 8 + 0
h. 300 + 8
j. 3 + 8 + 0
k. NH

3. What means the same as 948?

a. 9 + 4 + 8
b. 900 + 40 + 8
c. 900 + 4 + 8
d. 900 + 40 + 80
e. NH

4. What means the same as 807?

f. 80 + 7
g. 800 + 7
h. 8 + 0 + 7
j. 800 + 70
k. NH

5. What means the same as 355?

a. 3 + 5 + 5
b. 35 + 5
c. 300 + 550
d. 300 + 50 + 5
e. NH

6. What means the same as 6,702?

f. 6,000 + 70 + 2
g. 6,000 + 700 + 2
h. 6,000 + 700 + 20
j. 6 + 7 + 0 + 2
k. NH

7. What means the same as 8,076?

a. 8,000 + 70 + 6
b. 8,000 + 700 + 6
c. 8 + 0 + 7 + 6
d. 8,000 + 700 + 60
e. NH

8. What means the same as 9,140?

f. 9 + 1 + 4 + 0
g. 9,000 + 10 + 4
h. 9,000 + 100 + 40
j. 9,000 + 100 + 4
k. NH

Name _____   Skill: Place Value

Directions
Read each question and choose the correct answer. Mark the space for the answer you have chosen. Mark NH if the answer is not here.

1. What is the value of 5 in 758? a. ones b. tens c. hundreds d. thousands e. NH	5. What is the value of 4 in 114,256? a. hundred thousands b. ten thousands c. thousands d. hundreds e. NH
2. What is the value of 7 in 7,034? f. ones g. tens h. hundreds j. thousands k. NH	6. What is the value of 9 in 145,923? f. hundred thousands g. ten thousands h. thousands j. hundreds k. NH
3. What is the value of 1 in 21,400? a. ones b. hundreds c. thousands d. ten thousands e. NH	7. What is the value of 3 in 1,346,970? a. hundred thousands b. ten thousands c. thousands d. hundreds e. NH
4. What is the value of 8 in 69,782? f. ten thousands g. thousands h. hundreds j. tens k. NH	8. What is the value of 2 in 5,429,016? f. hundred thousands g. ten thousands h. thousands j. hundreds k. NH

© Carson-Dellosa CD-3754

Name _____   Skill: Place Value

Directions
Read each question and choose the correct answer. Mark the space for the answer you have chosen. Mark NH if the answer is not here.

1. What number has the greatest value? a. 798 b. 897 c. 789 d. 978 e. NH	5. What number has the smallest value? a. 530 b. 350 c. 305 d. 503 e. NH
2. What number has the greatest value? f. 1,564 g. 1,456 h. 1,465 j. 1,546 k. NH	6. What number has the smallest value? f. 6,423 g. 6,243 h. 6,234 j. 6,432 k. NH
3. What number has the greatest value? a. 76,109 b. 67,910 c. 76,901 d. 76,091 e. NH	7. What number has the smallest value? a. 80,483 b. 80,348 c. 83,038 d. 80,384 e. NH
4. What number has the greatest value? f. 50,439 g. 50,943 h. 53,940 j. 53,490 k. NH	8. What number has the smallest value? f. 4,102,800 g. 4,021,080 h. 4,012,800 j. 4,012,008 k. NH

© Carson-Dellosa CD-3754

Name _____

Skill: Place Value

Directions
Read each question and choose the correct answer. Mark the space for the answer you have chosen. Mark NH if the answer is not here.

---

1. What group is in order from least to greatest?

   a. 420   402   240
   b. 402   240   420
   c. 240   420   402
   d. 240   402   420
   e. NH

5. What group is in order from greatest to least?

   a. 766   676   667
   b. 667   676   766
   c. 766   667   676
   d. 667   766   676
   e. NH

---

2. What group is in order from least to greatest?

   f. 1,509   1,059   1,905
   g. 1,059   1,509   1,905
   h. 1,059   1,905   1,509
   j. 1,905   1,509   1,059
   k. NH

6. What group is in order from greatest to least?

   f. 5,959   5,599   5,995
   g. 5,599   5,959   5,995
   h. 5,995   5,959   5,599
   j. 5,995   5,599   5,959
   k. NH

---

3. What group is in order from least to greatest?

   a. 1,101   2,101   2,110
   b. 2,110   2,101   1,102
   c. 2,101   1,102   2,110
   d. 1,101   2,110   2,101
   e. NH

7. What group is in order from greatest to least?

   a. 10,101   10,011   11,001
   b. 11,001   10,011   10,101
   c. 10,011   10,101   11,001
   d. 11,001   10,101   10,011
   e. NH

---

4. What group is in order from least to greatest?

   f. 11,954   11,549   11,459
   g. 11,549   11,459   11,954
   h. 11,459   11,549   11,954
   j. 11,459   11,954   11,549
   k. NH

8. What group is in order from greatest to least?

   f. 91,500   90,150   95,100
   g. 95,100   90,150   91,500
   h. 95,100   91,500   90,150
   j. 90,150   91,500   95,100
   k. NH

© Carson-Dellosa CD-3754

Name _____   Skill: Fractions

Directions
Read each question and choose the correct answer. Mark the space for the answer you have chosen. Mark NH if the answer is not here.

---

1. What fraction is shaded?

a. $\frac{1}{2}$ d. $\frac{3}{4}$
b. $\frac{1}{3}$ e. NH
c. $\frac{1}{3}$

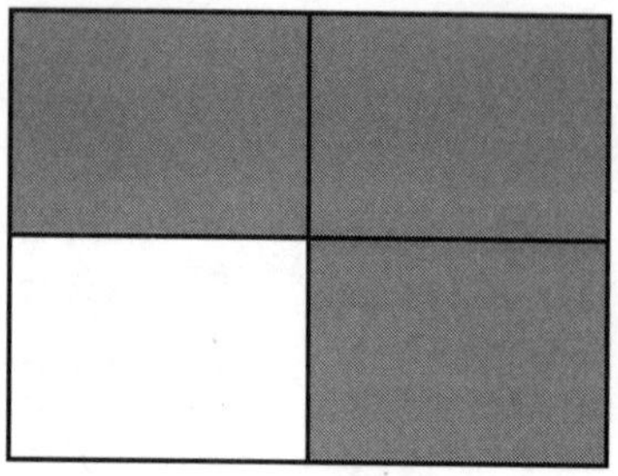

5. What fraction is shaded?

a. $3\frac{1}{4}$ d. $2\frac{3}{4}$
b. $2\frac{1}{3}$ e. NH
c. $2\frac{1}{4}$

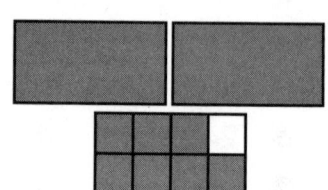

---

2. What fraction is shaded?

f. $\frac{1}{3}$ j. $\frac{1}{2}$
g. $\frac{3}{4}$ k. NH
h. $\frac{1}{4}$

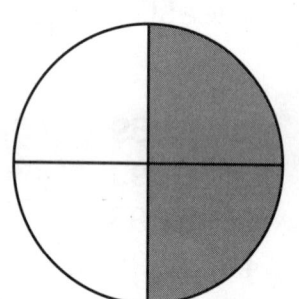

6. What fraction is shaded?

f. $1\frac{5}{6}$ j. $1\frac{2}{3}$
g. $1\frac{1}{6}$ k. NH
h. $1\frac{1}{5}$

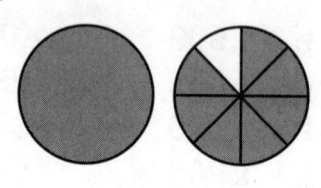

---

3. What fraction is shaded?

a. $\frac{2}{3}$ d. $\frac{1}{3}$
b. $\frac{1}{2}$ e. NH
c. $\frac{1}{4}$

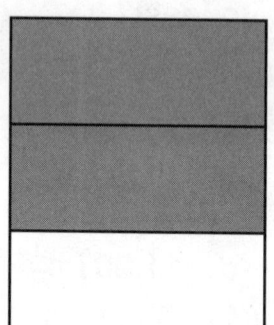

7. What fraction is shaded?

a. $2\frac{1}{2}$ d. $2\frac{1}{3}$
b. $2\frac{2}{3}$ e. NH
c. $2\frac{1}{4}$

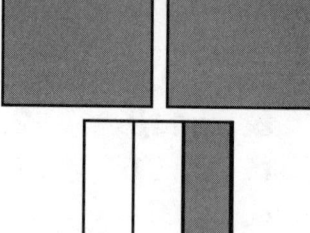

---

4. What fraction is shaded?

f. $\frac{7}{12}$ j. $\frac{5}{10}$
g. $\frac{5}{12}$ k. NH
h. $\frac{5}{7}$

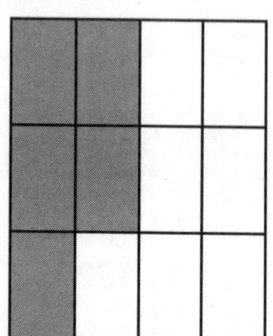

8. What fraction is shaded?

f. $1\frac{1}{7}$ j. $1\frac{7}{8}$
g. $1\frac{1}{8}$ k. NH
h. $2\frac{1}{8}$

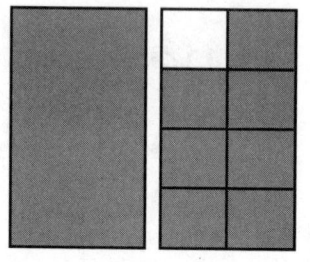

© Carson-Dellosa CD-3754

Name _____   Skill: Fractions

Directions
Read each question and choose the correct answer. Mark the space for the answer you have chosen. Mark NH if the answer is not here.

1. What fraction is greater than $\frac{1}{2}$?

   a. $\frac{2}{5}$ b. $\frac{2}{6}$ c. $\frac{5}{6}$ d. $\frac{1}{4}$ e. NH

2. What fraction is greater than $\frac{3}{4}$?

   f. $\frac{8}{9}$ g. $\frac{3}{10}$ h. $\frac{5}{9}$ j. $\frac{3}{5}$ k. NH

3. What fraction is greater than $\frac{7}{8}$?

   a. $\frac{4}{5}$ b. $\frac{3}{4}$ c. $\frac{7}{10}$ d. $\frac{9}{10}$ e. NH

4. What fraction is greater than $\frac{4}{5}$?

   f. $\frac{5}{6}$ g. $\frac{3}{4}$ h. $\frac{2}{3}$ j. $\frac{1}{2}$ k. NH

5. What fraction is less than $\frac{1}{3}$?

   a. $\frac{3}{9}$ b. $\frac{7}{10}$ c. $\frac{7}{1}$ d. $\frac{3}{10}$ e. NH

6. What fraction is less than $\frac{5}{9}$?

   f. $\frac{7}{10}$ g. $\frac{1}{4}$ h. $\frac{6}{8}$ j. $\frac{3}{4}$ k. NH

7. What fraction is less than $\frac{1}{2}$?

   a. $\frac{4}{7}$ b. $\frac{5}{8}$ c. $\frac{6}{7}$ d. $\frac{1}{9}$ e. NH

8. What fraction is less than $\frac{3}{8}$?

   f. $\frac{8}{9}$ g. $\frac{1}{4}$ h. $\frac{3}{5}$ j. $\frac{3}{4}$ k. NH

© Carson-Dellosa CD-3754

Name _____

Skill: Fractions

Directions
Read each question and choose the correct answer. Mark the space for the answer you have chosen. Mark NH if the answer is not here.

---

1. Which figure shows $\frac{1}{2}$ shaded?

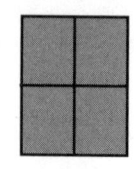 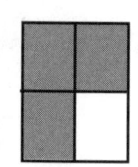 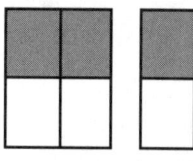 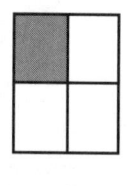

a.   b.   c.   d.   NH e.

5. Which figure shows $\frac{1}{2}$ shaded?

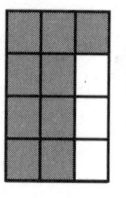 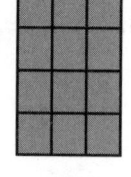 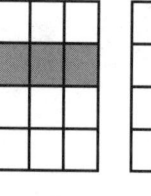

a.   b.   c.   d.   NH e.

---

2. Which figure shows $\frac{2}{3}$ shaded?

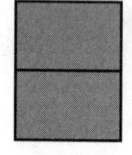 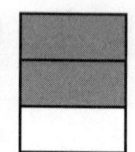 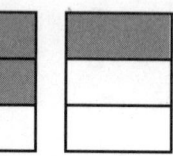 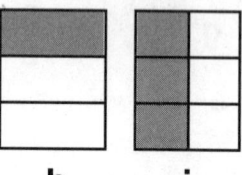 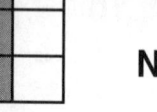

f.   g.   h.   j.   NH k.

6. Which figure shows $\frac{1}{7}$ shaded?

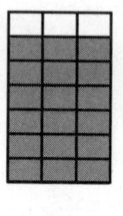 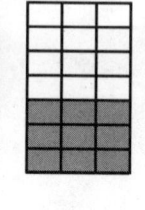 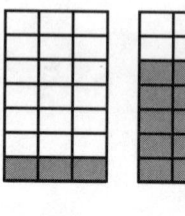

f.   g.   h.   j.   NH k.

---

3. Which figure shows $\frac{3}{4}$ shaded?

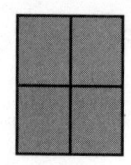 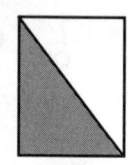 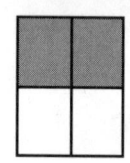 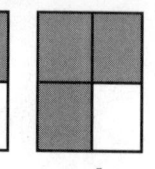

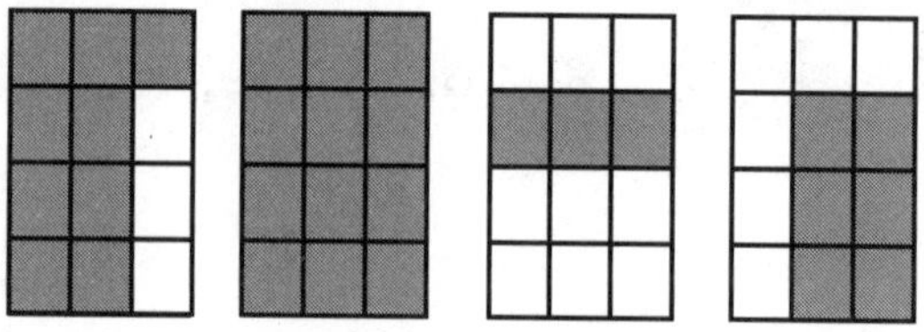

a.   b.   c.   d.   NH e.

7. Which figure shows $\frac{3}{5}$ shaded?

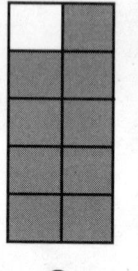 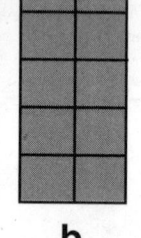 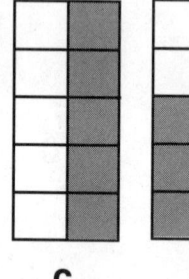

a.   b.   c.   d.   NH e.

---

4. Which figure shows $\frac{4}{6}$ shaded?

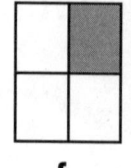 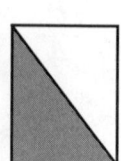 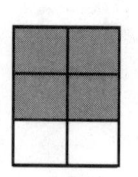 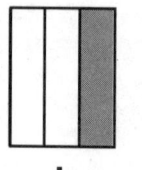

f.   g.   h.   j.   NH k.

8. Which figure shows $\frac{4}{8}$ shaded?

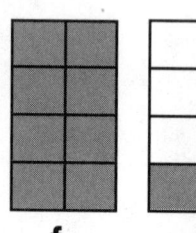 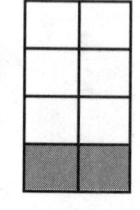 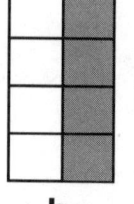 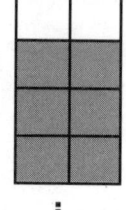

f.   g.   h.   j.   NH k.

© Carson-Dellosa CD-3754

Name _____   Skill: Operational Signs

Directions
Read each question and choose the correct answer. Mark the space for the answer you have chosen. Mark NH if the answer is not here.

1. What number sentence is true? a. 50 = 10 x 40 b. 50 = 10 – 40 c. 50 = 10 + 40 d. 50 = 10 ÷ 40 e. NH	**5.** What number sentence is true? a. 12 = 6 x 3 b. 12 = 6 – 3 c. 12 = 6 + 3 d. 12 = 6 ÷ 3 e. NH
2. What number sentence is true? f. 32 = 8 x 4 g. 32 = 8 ÷ 4 h. 32 = 8 + 4 j. 32 = 8 – 4 k. NH	**6.** What number sentence is true? f. 60 = 12 + 5 g. 60 = 12 – 5 h. 60 = 12 ÷ 5 j. 60 = 12 x 5 k. NH
3. What number sentence is true? a. 24 = 6 ÷ 4 b. 24 = 6 x 4 c. 24 = 6 – 4 d. 24 = 6 + 4 e. NH	**7.** What number sentence is true? a. 40 = 8 ÷ 5 b. 40 = 8 x 5 c. 40 = 8 – 5 d. 40 = 8 + 5 e. NH
4. What number sentence is true? f. 45 = 40 – 5 g. 45 = 40 ÷ 5 h. 45 = 40 + 5 j. 45 = 40 x 5 k. NH	**8.** What number sentence is true? f. 72 = 8 ÷ 9 g. 72 = 8 x 9 h. 72 = 8 – 9 j. 72 = 8 + 9 k. NH

© Carson-Dellosa CD-3754

Name _____ Skill: Operational Signs

Directions
Read each question and choose the correct answer. Mark the space for the answer you have chosen. Mark NH if the answer is not here.

---

1. What sign belongs in this statement?

    64 ◯ 8 = 8

    a. +
    b. −
    c. ÷
    d. >
    e. NH

2. What sign belongs in this statement?

    36 ◯ 12 x 3

    f. +
    g. x
    h. ÷
    j. =
    k. NH

3. What sign belongs in this statement?

    25 ◯ 4 = 100

    a. +
    b. =
    c. x
    d. ÷
    e. NH

4. What sign belongs in this statement?

    75 ◯ 15 = 60

    f. −
    g. +
    h. x
    j. ÷
    k. NH

5. What sign belongs in this statement?

    121 ◯ 11 = 11

    a. −
    b. x
    c. +
    d. ÷
    e. NH

6. What sign belongs in this statement?

    9 ◯ 7 = 63

    f. x
    g. ÷
    h. =
    j. >
    k. NH

7. What sign belongs in this statement?

    67 ◯ 17 = 50

    a. <
    b. +
    c. −
    d. ÷
    e. NH

8. What sign belongs in this statement?

    30 ◯ 3 = 90

    f. =
    g. +
    h. >
    j. x
    k. NH

© Carson-Dellosa CD-3754

Name _____   Skill: Greater Than, Less Than, Equal To

Directions
Read each question and choose the correct answer. Mark the space for the answer you have chosen. Mark NH if the answer is not here.

---

1. What sign belongs in this statement?

5 x 9 ◯ 30 + 5

a. >
b. <
c. ÷
d. =
e. NH

---

2. What sign belongs in this statement?

10 x 5 ◯ 60 − 20

f. +
g. >
h. +
j. <
k. NH

---

3. What sign belongs in this statement?

4 x 7 ◯ 50 − 20

a. >
b. =
c. <
d. x
e. NH

---

4. What sign belongs in this statement?

6 x 7 ◯ 21 + 21

f. >
g. −
h. <
j. =
k. NH

---

5. What sign belongs in this statement?

3 x 6 ◯ 2 x 9

a. +
b. =
c. <
d. >
e. NH

---

6. What sign belongs in this statement?

35 ÷ 7 ◯ 9 − 3

f. ÷
g. >
h. =
j. <
k. NH

---

7. What sign belongs in this statement?

5 x 6 ◯ 23 + 6

a. >
b. <
c. +
d. =
e. NH

---

8. What sign belongs in this statement?

50 + 12 ◯ 9 x 7

f. >
g. <
h. ÷
j. =
k. NH

© Carson-Dellosa CD-3754   16

Name _____   Skill: Mathematical Expressions

Directions
Read each question and choose the correct answer. Mark the space for the answer you have chosen. Mark NH if the answer is not here.

1. What has the same value as
7 + 7 + 7 + 7?

a. 14 – 7
b. 7 x 7 x 7 x 7
c. 4 x 7
d. 28 – 7
e. NH

2. What has the same value as 3 x 5?

f. 5 + 5 + 5
g. 3 + 3 + 3
h. 5 – 3
j. 5 + 3
k. NH

3. What has the same value as
40 + 50?

a. 40 x 50
b. 50 + 40
c. 50 – 40
d. 50 ÷ 40
e. NH

4. What has the same value as 6 x 10?

f. 10 x 10 x 10 x 10 x 10 x 10
g. 10 – 6
h. 10 + 6
j. 10 + 10 + 10 + 10 + 10 + 10
k. NH

5. What has the same value as
8 + 8 + 8?

a. 8 x 8 x 8
b. 3 + 8
c. 24 – 8
d. 8 x 3
e. NH

6. What has the same value as 2 x 10?

f. 10 x 10
g. 30 – 20
h. 20 ÷ 2
j. 5 x 4
k. NH

7. What has the same value as
5 + 5 + 5 + 5?

a. 5 x 5 x 5 x 5
b. 4 x 5
c. 20 – 5
d. 10 ÷ 10
e. NH

8. What has the same value as
25 + 10?

f. 30 + 15
g. 35 – 10
h. 35 x 0
j. 35 x 1
k. NH

© Carson-Dellosa CD-3754

Name _____   Skill: Mathematical Expressions

Directions
Read each question and choose the correct answer. Mark the space for the answer you have chosen. Mark NH if the answer is not here.

1. What has the same value as
   $40 \div 4$?

   a. $2 + 2 + 2 + 2 + 2$
   b. 5×5
   c. $12 - 1$
   d. $5 - 1$
   e. NH

2. What has the same value as
   $70 - 40$?

   f. $35 + 5$
   g. 6×5
   h. 5×5
   j. $20 + 15$
   k. NH

3. What has the same value as
   6×8?

   a. 4×12
   b. 5×12
   c. $50 - 9$
   d. $30 + 25$
   e. NH

4. What has the same value as
   $200 - 100$?

   f. $25 + 25 + 25$
   g. $75 + 75$
   h. $20 + 20 + 20 + 20 + 20$
   j. 5×25
   k. NH

5. What has the same value as
   $54 \div 9$?

   a. 2×4
   b. $10 - 5$
   c. $3 + 3$
   d. $4 + 3$
   e. NH

6. What has the same value as
   6×4?

   f. $12 + 10$
   g. $35 - 15$
   h. 5×5
   j. $8 + 8 + 8$
   k. NH

7. What has the same value as
   6×3?

   a. $3 \times 3 \times 3 \times 3 \times 3 \times 3$
   b. $6 + 6 + 6$
   c. $3 \times 3 \times 3$
   d. $20 - 3$
   e. NH

8. What has the same value as
   $25 + 10$?

   f. $4 \times 4 \times 4 \times 4$
   g. $7 + 7 + 7 + 7 + 7$
   h. $8 - 4$
   j. $4 + 4 + 4 + 4 + 4 + 4 + 4 + 4$
   k. NH

© Carson-Dellosa CD-3754

Name _____   Skill: Mathematical Expressions

Directions
Read each question and choose the correct answer. Mark the space for the answer you have chosen. Mark NH if the answer is not here.

1. What is the missing number?

    $18 - \square = 9$

    a. 7
    b. 17
    c. 2
    d. 8
    e. NH

2. What is the missing number?

    $100 - \square = 25$

    f. 75
    g. 25
    h. 50
    j. 20
    k. NH

3. What is the missing number?

    $8 + \square = 16$

    a. 5
    b. 25
    c. 8
    d. 7
    e. NH

4. What is the missing number?

    $100 - \square = 80$

    f. 1
    g. 80
    h. 10
    j. 20
    k. NH

5. What is the missing number?

    $\square \div 10 = 5$

    a. 55
    b. 5
    c. 50
    d. 60
    e. NH

6. What is the missing number?

    $\square \times 20 = 100$

    f. 4
    g. 5
    h. 6
    j. 7
    k. NH

7. What is the missing number?

    $40 \div \square = 8$

    a. 5
    b. 6
    c. 7
    d. 8
    e. NH

8. What is the missing number?

    $60 + \square = 100$

    f. 20
    g. 180
    h. 40
    j. 15
    k. NH

© Carson-Dellosa CD-3754

Name _____   Skill: Mathematical Expressions

Directions
Read each question and choose the correct answer. Mark the space for the answer you have chosen. Mark NH if the answer is not here.

1. What is the missing number?

$$5 + 7 + 8 = 8 + \square$$

- a. 5
- b. 7
- c. 10
- d. 12
- e. NH

2. What is the missing number?

$$7 + 7 + 3 = 10 + \square$$

- f. 3
- g. 7
- h. 8
- j. 14
- k. NH

3. What is the missing number?

$$9 + 8 + 7 = 2 \times \square$$

- a. 12
- b. 8
- c. 11
- d. 9
- e. NH

4. What is the missing number?

$$6 + 6 + 8 = 5 + \square$$

- f. 12
- g. 13
- h. 14
- j. 15
- k. NH

5. What is the missing number?

$$9 + 4 + 1 = 7 + \square$$

- a. 6
- b. 7
- c. 8
- d. 9
- e. NH

6. What is the missing number?

$$3 + 7 + 8 = 9 \times \square$$

- f. 6
- g. 5
- h. 4
- j. 3
- k. NH

7. What is the missing number?

$$9 + 9 + 7 = 5 \times \square$$

- a. 9
- b. 7
- c. 5
- d. 4
- e. NH

8. What is the missing number?

$$9 + 4 + 8 = 7 + 7 + \square$$

- f. 9
- g. 8
- h. 7
- j. 4
- k. NH

© Carson-Dellosa CD-3754

Name _____   Skill: Mathematical Expressions

Directions
Read each question and choose the correct answer. Mark the space for the answer you have chosen. Mark NH if the answer is not here.

1. What is the missing number?

$$4 \times \square = 20$$

a. 4
b. 5
c. 6
d. 7
e. NH

2. What is the missing number?

$$5 \times \square = 30$$

f. 3
g. 4
h. 5
j. 6
k. NH

3. What is the missing number?

$$10 \times \square = 30$$

a. 2
b. 3
c. 4
d. 5
e. NH

4. What is the missing number?

$$8 \times \square = 64$$

f. 6
g. 7
h. 8
j. 9
k. NH

5. What is the missing number?

$$3 \times \square = 18$$

a. 4
b. 5
c. 6
d. 7
e. NH

6. What is the missing number?

$$12 \times \square = 144$$

f. 9
g. 10
h. 11
j. 12
k. NH

7. What is the missing number?

$$5 \times \square = 35$$

a. 7
b. 8
c. 9
d. 10
e. NH

8. What is the missing number?

$$12 \times \square = 60$$

f. 3
g. 4
h. 5
j. 6
k. NH

© Carson-Dellosa CD-3754

Name _____

Skill: Mathematical Expressions

Directions
Read each question and choose the correct answer. Mark the space for the answer you have chosen. Mark NH if the answer is not here.

1. What is the missing number?

$$15 \div \square = 3$$

- a. 4
- b. 3
- c. 5
- d. 6
- e. NH

2. What is the missing number?

$$56 \div \square = 7$$

- f. 5
- g. 6
- h. 7
- j. 8
- k. NH

3. What is the missing number?

$$72 \div \square = 8$$

- a. 8
- b. 9
- c. 10
- d. 11
- e. NH

4. What is the missing number?

$$48 \div \square = 4$$

- f. 9
- g. 10
- h. 11
- j. 12
- k. NH

5. What is the missing number?

$$24 \div \square = 3$$

- a. 8
- b. 9
- c. 7
- d. 6
- e. NH

6. What is the missing number?

$$36 \div \square = 6$$

- f. 5
- g. 6
- h. 7
- j. 8
- k. NH

7. What is the missing number?

$$70 \div \square = 10$$

- a. 4
- b. 5
- c. 6
- d. 7
- e. NH

8. What is the missing number?

$$49 \div \square = 7$$

- f. 8
- g. 5
- h. 7
- j. 6
- k. NH

© Carson-Dellosa CD-3754

Name _____   Skill: Fact Families

Directions
Read each question and choose the correct answer. Mark the space for the answer you have chosen. Mark NH if the answer is not here.

1. Which number sentence is a member of this fact family?

    9 + 8 = 17

    a. 17 + 9 = 26
    b. 17 − 9 = 8
    c. 25 − 7 = 8
    d. 17 − 4 = 13
    e. NH

2. Which number sentence is a member of this fact family?

    2 x 4 = 8

    f. 2 x 8 = 16
    g. 8 ÷ 2 = 4
    h. 4 x 8 = 32
    j. 4 ÷ 2 = 2
    k. NH

3. Which number sentence is a member of this fact family?

    6 + 8 = 14

    a. 14 + 6 = 20
    b. 8 + 14 = 22
    c. 8 − 6 = 2
    d. 14 − 2 = 12
    e. NH

4. Which number sentence is a member of this fact family?

    28 ÷ 7 = 4

    f. 4 x 7 = 28
    g. 28 − 7 = 21
    h. 7 + 3 = 10
    j. 7 − 3 = 4
    k. NH

5. Which number sentence is a member of this fact family?

    12 − 7 = 5

    a. 17 − 12 = 5
    b. 7 − 5 = 2
    c. 12 + 7 = 19
    d. 5 + 7 = 12
    e. NH

6. Which number sentence is a member of this fact family?

    56 ÷ 8 = 7

    f. 7 + 8 = 15
    g. 7 x 8 = 56
    h. 56 ÷ 2 = 13
    j. 56 + 8 = 64
    k. NH

7. Which number sentence is a member of this fact family?

    13 − 9 = 4

    a. 13 + 4 = 17
    b. 17 − 13 = 4
    c. 9 + 4 = 13
    d. 9 − 4 = 5
    e. NH

8. Which number sentence is a member of this fact family?

    4 x 9 = 36

    f. 36 + 4 = 40
    g. 36 − 4 = 32
    h. 9 − 4 = 5
    j. 36 ÷ 9 = 4
    k. NH

© Carson-Dellosa CD-3754

Name _____   Skill: Fact Families

Directions
Read each question and choose the correct answer. Mark the space for the answer you have chosen. Mark NH if the answer is not here.

1. Which number sentence is a member of this fact family?

   60 ÷ 5 = 12

   a. 60 – 5 = 55
   b. 5 + 12 = 17
   c. 12 x 5 = 60
   d. 60 + 12 = 72
   e. NH

2. Which number sentence is a member of this fact family?

   11 – 4 = 7

   f. 11 + 7 = 18
   g. 11 + 4 = 15
   h. 18 – 7 = 11
   j. 15 – 4 = 11
   k. NH

3. Which number sentence is a member of this fact family?

   6 x 7 = 42

   a. 7 x 6 = 42
   b. 7 + 6 = 13
   c. 42 ÷ 2 = 21
   d. 42 + 7 = 49
   e. NH

4. Which number sentence is a member of this fact family?

   16 – 8 = 8

   f. 8 – 8 = 0
   g. 16 + 8 = 24
   h. 8 x 8 = 64
   j. 8 + 8 = 16
   k. NH

5. Which number sentence is a member of this fact family?

   10 x 8 = 80

   a. 80 – 8 = 72
   b. 10 – 8 = 2
   c. 80 – 10 = 70
   d. 80 ÷ 10 = 8
   e. NH

6. Which number sentence is a member of this fact family?

   4 + 6 = 10

   f. 10 – 6 = 4
   g. 10 + 6 = 16
   h. 16 – 6 = 10
   j. 4 x 6 = 24
   k. NH

7. Which number sentence is a member of this fact family?

   99 ÷ 9 = 11

   a. 99 ÷ 11 = 9
   b. 11 + 9 = 20
   c. 99 – 9 = 90
   d. 99 + 9 = 108
   e. NH

8. Which number sentence is a member of this fact family?

   3 x 4 = 12

   f. 4 – 3 = 1
   g. 12 – 4 = 8
   h. 12 ÷ 4 = 3
   j. 12 + 3 = 15
   k. NH

© Carson-Dellosa CD-3754

Name _____   Skill: Number Concepts Practice

Directions
Read each question and choose the correct answer. Mark the space for the answer you have chosen. Mark NH if the answer is not here.

1. What is the numeral for seventy-five thousand, four hundred?

    a. 75,004
    b. 75,040
    c. 75,400
    d. 750,400
    e. NH

2. What is the numeral for one hundred eight thousand, four hundred thirty-six?

    f. 108,436
    g. 184,036
    h. 184,360
    j. 180,436
    k. NH

3. What is the numeral for ten million, two hundred five thousand, one hundred eleven?

    a. 10,205,111
    b. 100,205,111
    c. 10,000,205,111
    d. 102,050,111
    e. NH

4. What is the missing number?

    | 800 | 850 | 900 | 950 | |

    f. 750
    g. 951
    h. 995
    j. 1,000
    k. NH

5. What number comes just <u>after</u> 199,999?

    a. 2,000,000
    b. 200,000
    c. 199,100
    d. 199,900
    e. NH

6. What number comes between 28,070 and 31,700?

    f. 27,999
    g. 31,701
    h. 28,090
    j. 32,650
    k. NH

7. What means the same as 654?

    a. 6 hundreds, 5 tens, and 0 ones
    b. 6 tens and 5 ones
    c. 6 hundreds, 6 tens, and 4 ones
    d. 6 hundreds, 4 tens, and 14 ones
    e. NH

8. What means the same as 7,392?

    f. 7,000 + 30 + 90 + 2
    g. 7,000 + 30 + 92
    h. 7,000 + 300 + 90 + 2
    j. 7 + 3 + 9 + 2
    k. NH

© Carson-Dellosa CD-3754

Name _____   Skill: Number Concepts Practice

Directions
Read each question and choose the correct answer. Mark the space for the answer you have chosen. Mark NH if the answer is not here.

1. In what place value is the 0 in the number 1,045,238?

   a. hundred thousands
   b. ten thousands
   c. thousands
   d. hundreds
   e. NH

2. Which number has the greatest value?

   f. 134,006
   g. 134,060
   h. 134,600
   j. 136,400
   k. NH

3. Which group is in order from least to greatest?

   a. 5,592 5,952 5,295
   b. 5,295 5,952 5,592
   c. 5,295 5,592 5,952
   d. 5,592 5,295 5,952
   e. NH

4. What fraction is shaded?

   f. $\frac{1}{5}$ j. $\frac{5}{12}$
   g. $\frac{2}{5}$ k. NH
   h. $\frac{1}{8}$

   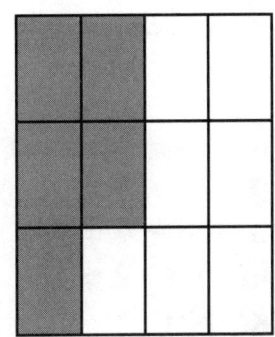

5. Which fraction is greater than $\frac{1}{3}$?

   a. $\frac{1}{5}$ d. $\frac{1}{4}$
   b. $\frac{1}{6}$ e. NH
   c. $\frac{1}{2}$

6. Which figure shows $\frac{6}{8}$?

   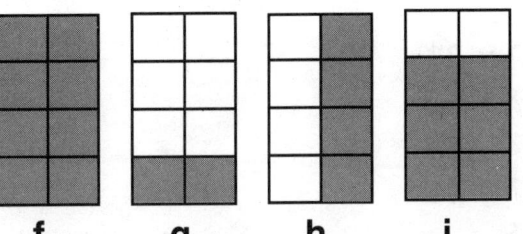

   f.   g.   h.   j.   NH   k.

7. Which number statement is true?

   a. $21 \div 3 > 7$
   b. $21 \div 3 = 7$
   c. $21 \times 3 < 7$
   d. $21 + 3 = 7$
   e. NH

8. What sign belongs in this statement?

   $8 \bigcirc 9 = 72$

   f. x
   g. +
   h. –
   j. ÷
   k. NH

© Carson-Dellosa CD-3754

Name _____   Skill: Number Concepts Practice

Directions
Read each question and choose the correct answer. Mark the space for the answer you have chosen. Mark NH if the answer is not here.

1. What sign makes this statement true?

   4 x 4 ◯ 20 – 6

   a. >
   b. –
   c. <
   d. =
   e. NH

2. What is another way of saying this?
   12 + 37

   f. 12 x 37
   g. 37 + 12
   h. 37 – 12
   j. 37 ÷ 12
   k. NH

3. What is another way of saying this?
   8 x 6

   a. 8 x 8 x 8 x 8 x 8 x 8
   b. 8 + 8 + 8 + 8
   c. 8 – 6
   d. 6 + 6 + 6 + 6 + 6 + 6 + 6 + 6
   e. NH

4. What is the missing number?

   42 – ☐ = 35

   f. 6
   g. 7
   h. 8
   j. 9
   k. NH

5. What is the missing number?

   3 + 5 + 9 = 7 + ☐

   a. 8
   b. 9
   c. 10
   d. 11
   e. NH

6. What is the missing number?

   6 x ☐ = 72

   f. 9
   g. 10
   h. 11
   j. 12
   k. NH

7. Which number is a member of this fact family?

   9 + 9 = 18

   a. 18 + 9 = 27
   b. 9 – 9 = 0
   c. 18 – 9 = 9
   d. 9 x 18 = 162
   e. NH

8. Which number sentence is a member of this fact family?

   4 x 12 = 48

   f. 48 – 12 = 36
   g. 12 x 4 = 48
   h. 48 – 4 = 44
   j. 4 + 12 = 16
   k. NH

© Carson-Dellosa CD-3754

Name _____   Skill: Number Concepts Practice

Directions
Read each question and choose the correct answer. Mark the space for the answer you have chosen. Mark NH if the answer is not here.

1. What is the numeral for one hundred nine thousand, twenty-nine?

   a. 109,209
   b. 190,209
   c. 109,029
   d. 109,290
   e. NH

2. What is the numeral for two hundred ninety-thousand, two hundred four?

   f. 200,090,204
   g. 290,204
   h. 209,204
   j. 2,090,240
   k. NH

3. What is the numeral for eleven million, one hundred fifteen thousand, one hundred?

   a. 11,100,015,100
   b. 11,115,000
   c. 11,101,510
   d. 11,115,100
   e. NH

4. What is the missing number?

   | 913 | 813 | 713 | 613 | |

   f. 500
   g. 513
   h. 612
   j. 600
   k. NH

5. What number comes just <u>before</u> 7,000?

   a. 7,001
   b. 6,000
   c. 6,999
   d. 6,900
   e. NH

6. What number comes between 12,090 and 14,180?

   f. 12,059
   g. 13,988
   h. 12,009
   j. 14,260
   k. NH

7. What means the same as 435?

   a. 4 hundreds, 5 tens, and 3 ones
   b. 5 hundreds, 3 tens, and 4 ones
   c. 4 hundreds, 2 tens, and 15 ones
   d. 4 hundreds, 3 tens, and 15 ones
   e. NH

8. What means the same as 8,960?

   f. 8,000 + 900 + 600
   g. 8,000 + 900 + 60
   h. 8,000 + 900 + 6
   j. 8 + 9 + 6 + 0
   k. NH

© Carson-Dellosa CD-3754

Name _____

Skill: Number Concepts Practice

Directions
Read each question and choose the correct answer. Mark the space for the answer you have chosen. Mark NH if the answer is not here.

1. In what place value is the 5 in the number 587,493?

   a. hundred thousands
   b. ten thousands
   c. thousands
   d. hundreds
   e. NH

2. Which number has the greatest value?

   f. 55,051
   g. 55,150
   h. 51,510
   j. 55,501
   k. NH

3. Which group is in order from greatest to least?

   a. 8,325   8,235   8,532
   b. 8,235   8,325   8,532
   c. 8,523   8,235   8,325
   d. 8,523   8,325   8,235
   e. NH

4. What fraction is shaded?

   f. $\frac{1}{2}$ j. $\frac{5}{6}$
   g. $\frac{1}{5}$ k. NH
   h. $\frac{1}{6}$

5. Which fraction is less than $\frac{1}{2}$?

   a. $\frac{3}{4}$ b. $\frac{3}{7}$ c. $\frac{4}{3}$ d. $\frac{4}{7}$ e. NH

6. Which fraction is less than $\frac{1}{3}$?

   f. $\frac{3}{5}$ g. $\frac{1}{2}$ h. $\frac{1}{4}$ j. $\frac{3}{8}$ k. NH

7. Which number statement is true?

   a. 45 x 9 = 5
   b. 45 > 9 x 5
   c. 45 = 9 x 5
   d. 45 − 9 = 5
   e. NH

8. What sign belongs in this statement?

   17 ◯ 9 = 8

   f. <
   g. +
   h. −
   j. ÷
   k. NH

© Carson-Dellosa CD-3754

Name _____   Skill: Number Concepts Practice

Directions
Read each question and choose the correct answer. Mark the space for the answer you have chosen. Mark NH if the answer is not here.

---

1. What sign makes this statement true?

   6 x 12 ◯ 9 x 9

   a. >
   b. −
   c. <
   d. =
   e. NH

---

2. What is another way of saying this?

   3 x 6

   f. 6 − 3
   g. 6 ÷ 3
   h. 6 x 3
   j. 6 + 3
   k. NH

---

3. What is another way of saying this?

   4 + 4 + 4 + 4 + 4 + 4

   a. 36 − 4
   b. 6 x 4
   c. 4 x 4 x 4 x 4 x 4 x 4
   d. 20 + 12
   e. NH

---

4. What is the missing number?

   45 + ☐ = 60

   f. 15
   g. 25
   h. 30
   j. 40
   k. NH

---

5. What is the missing number?

   2 + 7 + 7 = 7 + ☐

   a. 2
   b. 5
   c. 7
   d. 9
   e. NH

---

6. What is the missing number?

   6 x ☐ = 54

   f. 9
   g. 10
   h. 11
   j. 12
   k. NH

---

7. Which number sentence is a member of this fact family?

   16 − 9 = 7

   a. 9 − 7 = 1
   b. 16 + 7 = 23
   c. 16 − 7 = 9
   d. 16 + 9 = 25
   e. NH

---

8. Which number sentence is a member of this fact family?

   42 ÷ 6 = 7

   f. 42 + 7 = 49
   g. 42 − 6 = 36
   h. 42 ÷ 7 = 6
   j. 7 − 6 = 1
   k. NH

© Carson-Dellosa CD-3754

Name _____   Skill: Two-Digit Addition

Directions
Read each question and choose the correct answer. Mark the space for the answer you have chosen. Mark NH if the answer is not here.

1. $\begin{array}{r}55\\+\ 1\\\hline\end{array}$ a. 66 b. 54 c. 56 d. 551 e. NH	5. $\begin{array}{r}23\\+\ 5\\\hline\end{array}$ a. 22 b. 38 c. 28 d. 253 e. NH	9. $\begin{array}{r}43\\+24\\\hline\end{array}$ a. 21 b. 67 c. 61 d. 27 e. NH
2. $\begin{array}{r}41\\+\ 8\\\hline\end{array}$ f. 418 g. 47 h. 49 j. 48 k. NH	6. $\begin{array}{r}50\\+\ 7\\\hline\end{array}$ f. 12 g. 507 h. 58 j. 57 k. NH	10. $\begin{array}{r}52\\+24\\\hline\end{array}$ f. 36 g. 32 h. 72 j. 76 k. NH
3. $\begin{array}{r}82\\+\ 5\\\hline\end{array}$ a. 83 b. 87 c. 825 d. 86 e. NH	7. $\begin{array}{r}60\\+\ 8\\\hline\end{array}$ a. 68 b. 608 c. 70 d. 67 e. NH	11. $\begin{array}{r}18\\+40\\\hline\end{array}$ a. 50 b. 38 c. 30 d. 58 e. NH
4. $\begin{array}{r}28\\+\ 1\\\hline\end{array}$ f. 27 g. 28 h. 281 j. 29 k. NH	8. $\begin{array}{r}73\\+15\\\hline\end{array}$ f. 88 g. 62 h. 68 j. 82 k. NH	12. $\begin{array}{r}75\\+21\\\hline\end{array}$ f. 54 g. 96 h. 56 j. 94 k. NH

Name _____

Skill: Two-Digit Addition

Directions
Read each question and choose the correct answer. Mark the space for the answer you have chosen. Mark NH if the answer is not here.

1.
    61
  + 7

  a. 66
  b. 617
  c. 69
  d. 68
  e. NH

2.
    64
  + 4

  f. 67
  g. 60
  h. 644
  j. 68
  k. NH

3.
    33
  + 6

  a. 43
  b. 336
  c. 34
  d. 39
  e. NH

4.
    47
  + 12

  f. 39
  g. 58
  h. 59
  j. 55
  k. NH

5.
    46
  + 32

  a. 14
  b. 78
  c. 79
  d. 74
  e. NH

6.
    56
  + 13

  f. 69
  g. 43
  h. 49
  j. 63
  k. NH

7.
    57
  + 30

  a. 807
  b. 87
  c. 27
  d. 80
  e. NH

8.
    24
  + 45

  f. 21
  g. 29
  h. 61
  j. 69
  k. NH

9.
    32
  + 56

  a. 28
  b. 84
  c. 87
  d. 88
  e. NH

10.
    80
  + 17

  f. 817
  g. 97
  h. 77
  j. 90
  k. NH

11.
    53
  + 21

  a. 32
  b. 74
  c. 34
  d. 72
  e. NH

12.
    81
  + 16

  f. 97
  g. 75
  h. 77
  j. 95
  k. NH

© Carson-Dellosa CD-3754

Name _____   Skill: Two- and Three-Digit Addition

Directions
Read each question and choose the correct answer. Mark the space for the answer you have chosen. Mark NH if the answer is not here.

1. 471 + 24	a. 455 b. 453 c. 495 d. 485 e. NH	5. 817 + 81	a. 898 b. 876 c. 878 d. 897 e. NH	9. 704 + 291	a. 593 b. 995 c. 905 d. 993 e. NH
2. 729 + 20	f. 749 g. 709 h. 720 j. 740 k. NH	6. 665 + 24	f. 649 g. 689 h. 681 j. 641 k. NH	10. 530 + 327	f. 817 g. 867 h. 850 j. 857 k. NH
3. 263 + 15	a. 727 b. 278 c. 252 d. 258 e. NH	7. 371 + 106	a. 475 b. 476 c. 477 d. 577 e. NH	11. 254 + 320	a. 534 b. 570 c. 574 d. 575 e. NH
4. 403 + 64	f. 467 g. 461 h. 494 j. 441 k. NH	8. 843 + 123	f. 766 g. 960 h. 926 j. 966 k. NH	12. 332 + 426	f. 768 g. 758 h. 754 j. 757 k. NH

© Carson-Dellosa CD-3754

Name _____

Skill: Two- and Three-Digit Addition

Directions
Read each question and choose the correct answer. Mark the space for the answer you have chosen. Mark NH if the answer is not here.

1. 629 + 74 a. 693 b. 713 c. 705 d. 703 e. NH	5. 723 + 68 a. 781 b. 791 c. 891 d. 745 e. NH	9. 471 + 269 a. 740 b. 630 c. 730 d. 750 e. NH
2. 836 + 85 f. 921 g. 821 h. 831 j. 911 k. NH	6. 554 + 48 f. 592 g. 604 h. 602 j. 594 k. NH	10. 470 + 379 f. 749 g. 840 h. 849 j. 709 k. NH
3. 727 + 64 a. 791 b. 781 c. 751 d. 891 e. NH	7. 197 + 604 a. 791 b. 801 c. 811 d. 711 e. NH	11. 363 + 189 a. 442 b. 552 c. 452 d. 542 e. NH
4. 563 + 98 f. 440 g. 336 h. 436 j. 441 k. NH	8. 588 + 147 f. 621 g. 721 h. 731 j. 735 k. NH	12. 499 + 156 f. 545 g. 655 h. 645 j. 635 k. NH

© Carson-Dellosa CD-3754

Name _____   Skill: Three- and Four-Digit Addition

Directions
Read each question and choose the correct answer. Mark the space for the answer you have chosen. Mark NH if the answer is not here.

1. 4,733 + 244 a. 4,577 b. 5,977 c. 4,911 d. 4,971 e. NH	5. 6,371 + 408 a. 6,779 b. 7,179 c. 6,777 d. 6,709 e. NH	9. 7,132 + 1,622 a. 8,554 b. 8,710 c. 8,714 d. 8,754 e. NH
2. 3,560 + 129 f. 3,449 g. 3,689 h. 3,699 j. 3,680 k. NH	6. 1,154 + 513 f. 1,641 g. 1,661 h. 1,667 j. 1,677 k. NH	10. 2,214 + 3,674 f. 5,588 g. 5,788 h. 5,880 j. 5,888 k. NH
3. 8,304 + 253 a. 8,507 b. 8,558 c. 8,157 d. 8,657 e. NH	7. 1,231 + 5,665 a. 6,996 b. 6,794 c. 6,896 d. 6,894 e. NH	11. 6,131 + 1,722 a. 7,653 b. 7,853 c. 7,713 d. 7,953 e. NH
4. 9,207 + 630 f. 9,437 g. 9,807 h. 9,837 j. 9,830 k. NH	8. 6,704 + 1,103 f. 8,707 g. 7,607 h. 7,601 j. 7,807 k. NH	12. 5,471 + 1,323 f. 6,794 g. 5,654 h. 6,764 j. 6,894 k. NH

© Carson-Dellosa CD-3754

Name _____   Skill: Three- and Four-Digit Addition

Directions
Read each question and choose the correct answer. Mark the space for the answer you have chosen. Mark NH if the answer is not here.

1. 5,739 + 484	a. 5,223 b. 6,223 c. 6,123 d. 6,243 e. NH	5. 2,526 + 879	a. 3,415 b. 3,405 c. 3,305 d. 3,315 e. NH	9. 4,679 + 532	a. 5,211 b. 5,111 c. 4,211 d. 4,111 e. NH
2. 8,195 + 816	f. 8,901 g. 8,701 h. 9,001 j. 9,011 k. NH	6. 1,863 + 287	f. 1,040 g. 2,050 h. 2,150 j. 2,254 k. NH	10. 5,157 + 857	f. 5,904 g. 6,014 h. 5,914 j. 6,114 k. NH
3. 4,973 + 679	a. 5,552 b. 5,752 c. 5,652 d. 4,752 e. NH	7. 3,385 + 616	a. 3,371 b. 3,991 c. 3,901 d. 4,001 e. NH	11. 3,659 + 3,471	a. 6,028 b. 7,130 c. 6,130 d. 7,120 e. NH
4. 3,687 + 999	f. 4,576 g. 4,696 h. 4,686 j. 4,596 k. NH	8. 2,358 + 876	f. 3,134 g. 3,244 h. 3,234 j. 3,334 k. NH	12. 5,773 + 5,658	f. 11,431 g. 10,322 h. 10,331 j. 11,331 k. NH

© Carson-Dellosa CD-3754

Name _____

Skill: Addition with Three Addends

Directions
Read each question and choose the correct answer. Mark the space for the answer you have chosen. Mark NH if the answer is not here.

1. 43 10 + 32 a. 65 b. 75 c. 85 d. 71 e. NH	5. 31 30 + 35 a. 94 b. 96 c. 950 d. 86 e. NH	9. 24 31 + 22 a. 87 b. 57 c. 67 d. 77 e. NH
2. 12 51 + 10 f. 73 g. 83 h. 63 j. 71 k. NH	6. 41 16 + 22 f. 59 g. 89 h. 79 j. 88 k. NH	10. 35 41 + 22 f. 76 g. 97 h. 77 j. 87 k. NH
3. 60 13 + 12 a. 85 b. 45 c. 86 d. 84 e. NH	7. 21 35 + 31 a. 67 b. 97 c. 76 d. 87 e. NH	11. 14 22 + 41 a. 66 b. 76 c. 77 d. 78 e. NH
4. 5 43 + 30 f. 73 g. 72 h. 18 j. 78 k. NH	8. 11 40 + 6 f. 51 g. 56 h. 57 j. 67 k. NH	12. 14 30 + 24 f. 58 g. 68 h. 78 j. 88 k. NH

© Carson-Dellosa CD-3754

Name _____

Skill: Addition with Three Addends

Directions
Read each question and choose the correct answer. Mark the space for the answer you have chosen. Mark NH if the answer is not here.

1.

   428
    26
 + 41

a. 494
b. 495
c. 595
d. 493
e. NH

2.

   169
    75
 + 43

f. 187
g. 197
h. 287
j. 297
k. NH

3.

   547
    33
 + 51

a. 631
b. 641
c. 531
d. 541
e. NH

4.

   239
    84
 + 46

f. 269
g. 259
h. 369
j. 359
k. NH

5.

   235
    58
 + 81

a. 374
b. 375
c. 274
d. 275
e. NH

6.

   425
    86
 + 19

f. 519
g. 529
h. 620
j. 530
k. NH

7.

   813
    83
 + 18

a. 904
b. 924
c. 1,014
d. 914
e. NH

8.

   783
    49
 + 15

f. 837
g. 847
h. 836
j. 737
k. NH

9.

   404
    97
 + 23

a. 424
b. 524
c. 534
d. 536
e. NH

10.

   836
    52
 + 48

f. 826
g. 846
h. 946
j. 936
k. NH

11.

   571
   128
 + 26

a. 723
b. 724
c. 725
d. 726
e. NH

12.

   267
   371
 + 67

f. 605
g. 595
h. 705
j. 695
k. NH

© Carson-Dellosa CD-3754

Name _____   Skill: Two-Digit Subtraction

Directions
Read each question and choose the correct answer. Mark the space for the answer you have chosen. Mark NH if the answer is not here.

1.
36
− 3
a. 30
b. 39
c. 33
d. 32
e. NH

5.
48
− 4
a. 42
b. 44
c. 46
d. 48
e. NH

9.
87
− 32
a. 45
b. 55
c. 56
d. 57
e. NH

2.
59
− 5
f. 53
g. 52
h. 54
j. 55
k. NH

6.
78
− 2
f. 80
g. 78
h. 77
j. 76
k. NH

10.
66
− 14
f. 52
g. 72
h. 60
j. 53
k. NH

3.
42
− 1
a. 43
b. 31
c. 41
d. 33
e. NH

7.
87
− 21
a. 78
b. 68
c. 66
d. 65
e. NH

11.
98
− 40
a. 50
b. 51
c. 41
d. 40
e. NH

4.
49
− 7
f. 22
g. 36
h. 23
j. 21
k. NH

8.
78
− 55
f. 33
g. 32
h. 22
j. 23
k. NH

12.
95
− 33
f. 72
g. 62
h. 68
j. 78
k. NH

© Carson-Dellosa CD-3754   39

Name _____

Skill: Two-Digit Subtraction

Directions
Read each question and choose the correct answer. Mark the space for the answer you have chosen. Mark NH if the answer is not here.

1. 47 − 5 a. 42 b. 52 c. 44 d. 43 e. NH	5. 94 − 4 a. 98 b. 90 c. 91 d. 92 e. NH	9. 69 − 21 a. 48 b. 47 c. 46 d. 38 e. NH
2. 67 − 3 f. 70 g. 64 h. 65 j. 62 k. NH	6. 39 − 2 f. 41 g. 38 h. 37 j. 36 k. NH	10. 85 − 22 f. 67 g. 57 h. 63 j. 53 k. NH
3. 24 − 2 a. 20 b. 22 c. 23 d. 26 e. NH	7. 73 − 33 a. 43 b. 40 c. 46 d. 50 e. NH	11. 88 − 71 a. 19 b. 29 c. 27 d. 17 e. NH
4. 67 − 2 f. 69 g. 55 h. 59 j. 65 k. NH	8. 89 − 53 f. 32 g. 35 h. 36 j. 46 k. NH	12. 58 − 21 f. 39 g. 37 h. 27 j. 49 k. NH

© Carson-Dellosa CD-3754

Name _____   Skill: Two- and Three-Digit Subtraction

Directions
Read each question and choose the correct answer. Mark the space for the answer you have chosen. Mark NH if the answer is not here.

1. 456 − 31 a. 485 b. 445 c. 415 d. 425 e. NH	5. 856 − 40 a. 810 b. 826 c. 816 d. 896 e. NH	9. 983 − 471 a. 512 b. 452 c. 552 d. 524 e. NH
2. 788 − 46 f. 744 g. 742 h. 743 j. 752 k. NH	6. 576 − 43 f. 533 g. 543 h. 532 j. 539 k. NH	10. 598 − 33 f. 575 g. 576 h. 566 j. 565 k. NH
3. 729 − 18 a. 711 b. 731 c. 717 d. 712 e. NH	7. 789 − 729 a. 79 b. 760 c. 60 d. 50 e. NH	11. 487 − 106 a. 382 b. 281 c. 381 d. 302 e. NH
4. 786 − 52 f. 754 g. 724 h. 734 j. 735 k. NH	8. 825 − 324 f. 521 g. 409 h. 501 j. 509 k. NH	12. 678 − 533 f. 155 g. 145 h. 135 j. 45 k. NH

© Carson-Dellosa CD-3754

Name _____

Skill: Two- and Three-Digit Subtraction

Directions
Read each question and choose the correct answer. Mark the space for the answer you have chosen. Mark NH if the answer is not here.

1.

578
− 99

a. 479
b. 521
c. 567
d. 467
e. NH

5.

850
− 79

a. 781
b. 829
c. 771
d. 761
e. NH

9.

837
− 158

a. 689
b. 685
c. 721
d. 679
e. NH

2.

921
− 76

f. 945
g. 845
h. 955
j. 856
k. NH

6.

312
− 63

f. 351
g. 251
h. 349
j. 249
k. NH

10.

482
− 287

f. 195
g. 205
h. 185
j. 165
k. NH

3.

834
− 79

a. 755
b. 745
c. 845
d. 855
e. NH

7.

823
− 294

a. 671
b. 629
c. 529
d. 571
e. NH

11.

934
− 396

a. 638
b. 538
c. 528
d. 548
e. NH

4.

601
− 56

f. 545
g. 645
h. 555
j. 455
k. NH

8.

714
− 397

f. 417
g. 317
h. 483
j. 383
k. NH

12.

830
− 278

f. 648
g. 642
h. 552
j. 542
k. NH

© Carson-Dellosa CD-3754

Name _____   Skill: Three- and Four-Digit Subtraction

Directions
Read each question and choose the correct answer. Mark the space for the answer you have chosen. Mark NH if the answer is not here.

1. 5,706 − 502 a. 5,208 b. 5,204 c. 5,304 d. 4,204 e. NH	5. 6,852 − 410 a. 6,442 b. 6,542 c. 6,432 d. 6,440 e. NH	9. 4,685 − 1,361 a. 3,324 b. 3,224 c. 3,326 d. 3,924 e. NH
2. 9,835 − 312 f. 9,543 g. 9,523 h. 9,423 j. 7,523 k. NH	6. 5,789 − 723 f. 4,066 g. 5,066 h. 5,966 j. 5,063 k. NH	10. 9,837 − 7,215 f. 1,622 g. 1,722 h. 2,622 j. 2,522 k. NH
3. 4,793 − 520 a. 4,263 b. 4,353 c. 4,373 d. 4,273 e. NH	7. 9,870 − 1,470 a. 8,400 b. 8,500 c. 8,470 d. 8,570 e. NH	11. 7,487 − 6,305 a. 1,282 b. 1,182 c. 182 d. 1,782 e. NH
4. 6,379 − 243 f. 6,536 g. 6,236 h. 6,136 j. 6,036 k. NH	8. 8,457 − 2,141 f. 5,326 g. 6,316 h. 5,316 j. 6,318 k. NH	12. 9,778 − 8,523 f. 1,234 g. 1,245 h. 1,155 j. 1,255 k. NH

© Carson-Dellosa CD-3754

Name _____

Skill: Three- and Four-Digit Subtraction

Directions
Read each question and choose the correct answer. Mark the space for the answer you have chosen. Mark NH if the answer is not here.

1.
6,301
− 529

a. 6,228
b. 6,828
c. 5,772
d. 5,782
e. NH

5.
8,145
− 789

a. 7,356
b. 8,844
c. 7,456
d. 7,366
e. NH

9.
7,317
− 694

a. 7,623
b. 6,723
c. 9,933
d. 6,623
e. NH

2.
4,723
− 524

f. 3,199
g. 4,299
h. 4,199
j. 4,209
k. NH

6.
2,205
− 756

f. 2,549
g. 1,449
h. 1,559
j. 1,459
k. NH

10.
9,270
− 5,856

f. 3,414
g. 4,626
h. 3,514
j. 3,416
k. NH

3.
5,572
− 675

a. 4,897
b. 5,103
c. 4,887
d. 4,907
e. NH

7.
9,004
− 4,788

a. 4,226
b. 4,216
c. 4,326
d. 5,226
e. NH

11.
8,543
− 3,627

a. 5,124
b. 5,916
c. 4,916
d. 4,816
e. NH

4.
2,153
− 883

f. 1,270
g. 1,280
h. 1,380
j. 1,370
k. NH

8.
5,034
− 1,748

f. 4,314
g. 4,276
h. 3,286
j. 3,276
k. NH

12.
8,162
− 2,468

f. 5,684
g. 5,694
h. 5,794
j. 5,784
k. NH

© Carson-Dellosa CD-3754

Name _____

Directions
Read each question and choose the correct answer. Mark the space for the answer you have chosen. Mark NH if the answer is not here.

Skill: Multiplication (Facts to 12)

1. $\begin{array}{r}6\\ \times 2\\ \hline\end{array}$ a. 18 b. 15 c. 12 d. 9 e. NH	5. $\begin{array}{r}4\\ \times 5\\ \hline\end{array}$ a. 16 b. 18 c. 19 d. 20 e. NH	9. $\begin{array}{r}6\\ \times 3\\ \hline\end{array}$ a. 12 b. 18 c. 15 d. 24 e. NH
2. $\begin{array}{r}7\\ \times 3\\ \hline\end{array}$ f. 27 g. 11 h. 21 j. 24 k. NH	6. $\begin{array}{r}5\\ \times 0\\ \hline\end{array}$ f. 5 g. 0 h. 10 j. 1 k. NH	10. $\begin{array}{r}8\\ \times 6\\ \hline\end{array}$ f. 42 g. 56 h. 49 j. 48 k. NH
3. $\begin{array}{r}7\\ \times 4\\ \hline\end{array}$ a. 14 b. 28 c. 32 d. 35 e. NH	7. $\begin{array}{r}5\\ \times 9\\ \hline\end{array}$ a. 35 b. 30 c. 40 d. 45 e. NH	11. $\begin{array}{r}8\\ \times 3\\ \hline\end{array}$ a. 18 b. 22 c. 24 d. 27 e. NH
4. $\begin{array}{r}9\\ \times 6\\ \hline\end{array}$ f. 54 g. 56 h. 63 j. 48 k. NH	8. $\begin{array}{r}8\\ \times 8\\ \hline\end{array}$ f. 56 g. 62 h. 71 j. 64 k. NH	12. $\begin{array}{r}7\\ \times 6\\ \hline\end{array}$ f. 38 g. 40 h. 42 j. 48 k. NH

© Carson-Dellosa CD-3754

Name _____   Skill: Multiplication (Facts to 12)

Directions
Read each question and choose the correct answer. Mark the space for the answer you have chosen. Mark NH if the answer is not here.

1.

   10
  x 5

a. 105
b. 50
c. 150
d. 500
e. NH

2.

   9
  x 4

f. 36
g. 32
h. 35
j. 40
k. NH

3.

   9
  x 9

a. 81
b. 90
c. 72
d. 77
e. NH

4.

   11
  x 10

f. 1,110
g. 110
h. 100
j. 101
k. NH

5.

   8
  x 9

a. 81
b. 63
c. 72
d. 74
e. NH

6.

   11
  x 11

f. 111
g. 121
h. 1,111
j. 122
k. NH

7.

   12
  x 2

a. 22
b. 122
c. 26
d. 24
e. NH

8.

   9
  x 7

f. 84
g. 64
h. 63
j. 72
k. NH

9.

   11
  x 7

a. 9
b. 7
c. 5
d. 6
e. NH

10.

   7
  x 8

f. 72
g. 54
h. 63
j. 56
k. NH

11.

   7
  x 7

a. 54
b. 48
c. 49
d. 14
e. NH

12.

   12
  x 5

f. 72
g. 86
h. 60
j. 65
k. NH

Name _____

Directions

Skill: Multiplication (Facts to 12)

Read each question and choose the correct answer. Mark the space for the answer you have chosen. Mark NH if the answer is not here.

1. 9 x2	a. 18 b. 16 c. 10 d. 6 e. NH	5. 9 x6	a. 72 b. 56 c. 63 d. 54 e. NH	9. 8 x7	a. 54 b. 56 c. 63 d. 72 e. NH
2. 12 x0	f. 12 g. 120 h. 102 j. 0 k. NH	6. 5 x11	f. 55 g. 115 h. 511 j. 50 k. NH	10. 10 x10	f. 100 g. 10 h. 1,000 j. 110 k. NH
3. 12 x7	a. 60 b. 72 c. 74 d. 84 e. NH	7. 7 x6	a. 48 b. 46 c. 42 d. 13 e. NH	11. 4 x4	a. 44 b. 18 c. 16 d. 12 e. NH
4. 10 x3	f. 55 g. 30 h. 25 j. 10 k. NH	8. 11 x8	f. 118 g. 88 h. 81 j. 18 k. NH	12. 5 x5	f. 55 g. 30 h. 20 j. 25 k. NH

© Carson-Dellosa CD-3754

Name _____   Skill: Multiplication (Two-Digit Numbers)

Directions
Read each question and choose the correct answer. Mark the space for the answer you have chosen. Mark NH if the answer is not here.

1. 26 × 5 a. 130 b. 106 c. 120 d. 140 e. NH	5. 19 × 2 a. 218 b. 192 c. 28 d. 38 e. NH	9. 50 × 4 a. 2,000 b. 200 c. 540 d. 450 e. NH
2. 55 × 3 f. 175 g. 355 h. 155 j. 165 k. NH	6. 61 × 5 f. 355 g. 315 h. 305 j. 325 k. NH	10. 71 × 6 f. 421 g. 486 h. 426 j. 481 k. NH
3. 18 × 7 a. 85 b. 156 c. 126 d. 136 e. NH	7. 59 × 2 a. 118 b. 109 c. 119 d. 127 e. NH	11. 47 × 5 a. 207 b. 237 c. 245 d. 235 e. NH
4. 37 × 6 f. 122 g. 222 h. 232 j. 242 k. NH	8. 83 × 4 f. 252 g. 332 h. 232 j. 352 k. NH	12. 91 × 3 f. 271 g. 273 h. 913 j. 319 k. NH

© Carson-Dellosa CD-3754

Name _____   Skill: Multiplication (Two-Digit Numbers)

Directions
Read each question and choose the correct answer. Mark the space for the answer you have chosen. Mark NH if the answer is not here.

1.

86
x 4

a. 254
b. 244
c. 354
d. 344
e. NH

5.

56
x 3

a. 152
b. 158
c. 168
d. 172
e. NH

9.

90
x 6

a. 540
b. 560
c. 630
d. 54
e. NH

2.

20
x 5

f. 100
g. 80
h. 1000
j. 110
k. NH

6.

79
x 2

f. 148
g. 158
h. 168
j. 138
k. NH

10.

43
x 8

f. 324
g. 344
h. 364
j. 244
k. NH

3.

82
x 7

a. 574
b. 564
c. 572
d. 544
e. NH

7.

28
x 6

a. 128
b. 138
c. 158
d. 168
e. NH

11.

86
x 2

a. 172
b. 162
c. 192
d. 202
e. NH

4.

65
x 8

f. 480
g. 520
h. 425
j. 510
k. NH

8.

94
x 4

f. 366
g. 362
h. 376
j. 372
k. NH

12.

91
x 5

f. 451
g. 455
h. 465
j. 485
k. NH

© Carson-Dellosa CD-3754   49

Name _____ Skill: Multiplication (Two-Digit Numbers)

Directions
Read each question and choose the correct answer. Mark the space for the answer you have chosen. Mark NH if the answer is not here.

1. 72 × 6	a. 422 b. 412 c. 432 d. 442 e. NH	5. 45 × 8	a. 320 b. 335 c. 360 d. 365 e. NH	9. 29 × 7	a. 203 b. 214 c. 193 d. 183 e. NH
2. 58 × 4	f. 208 g. 232 h. 242 j. 254 k. NH	6. 70 × 9	f. 540 g. 560 h. 720 j. 630 k. NH	10. 64 × 5	f. 304 g. 314 h. 324 j. 320 k. NH
3. 79 × 8	a. 632 b. 562 c. 622 d. 612 e. NH	7. 56 × 3	a. 158 b. 168 c. 162 d. 148 e. NH	11. 75 × 7	a. 495 b. 505 c. 525 d. 535 e. NH
4. 83 × 6	f. 488 g. 478 h. 498 j. 508 k. NH	8. 47 × 5	f. 215 g. 235 h. 225 j. 230 k. NH	12. 39 × 9	f. 351 g. 217 h. 291 j. 341 k. NH

© Carson-Dellosa CD-3754

Name _____   Skill: Division (Facts to 12)

Directions
Read each question and choose the correct answer. Mark the space for the answer you have chosen. Mark NH if the answer is not here.

1. $3\overline{)21}$ a. 8 b. 7 c. 6 d. 5 e. NH	5. $6\overline{)42}$ a. 5 b. 6 c. 7 d. 8 e. NH	9. $2\overline{)6}$ a. 1 b. 2 c. 3 d. 4 e. NH
2. $2\overline{)18}$ f. 6 g. 7 h. 8 j. 9 k. NH	6. $8\overline{)32}$ f. 4 g. 5 h. 6 j. 7 k. NH	10. $4\overline{)28}$ f. 4 g. 5 h. 6 j. 7 k. NH
3. $5\overline{)35}$ a. 8 b. 7 c. 6 d. 5 e. NH	7. $4\overline{)16}$ a. 6 b. 5 c. 4 d. 3 e. NH	11. $3\overline{)24}$ a. 9 b. 8 c. 7 d. 6 e. NH
4. $8\overline{)32}$ f. 4 g. 5 h. 6 j. 7 k. NH	8. $6\overline{)24}$ f. 6 g. 5 h. 4 j. 3 k. NH	12. $7\overline{)49}$ f. 6 g. 7 h. 8 j. 9 k. NH

© Carson-Dellosa CD-3754

Name _____   Skill: Division (Facts to 12)

Directions
Read each question and choose the correct answer. Mark the space for the answer you have chosen. Mark NH if the answer is not here.

1. 6)‾42 a. 5 b. 6 c. 8 d. 7 e. NH	5. 9)‾54 a. 6 b. 8 c. 9 d. 7 e. NH	9. 8)‾48 a. 7 b. 8 c. 9 d. 10 e. NH
2. 5)‾25 f. 7 g. 6 h. 5 j. 4 k. NH	6. 9)‾72 f. 11 g. 10 h. 9 j. 8 k. NH	10. 6)‾60 f. 9 g. 10 h. 11 j. 12 k. NH
3. 7)‾49 a. 7 b. 3 c. 4 d. 6 e. NH	7. 9)‾81 a. 10 b. 9 c. 8 d. 7 e. NH	11. 6)‾36 a. 8 b. 6 c. 7 d. 5 e. NH
4. 6)‾48 f. 6 g. 7 h. 8 j. 9 k. NH	8. 5)‾45 f. 6 g. 7 h. 8 j. 9 k. NH	12. 4)‾32 f. 6 g. 7 h. 8 j. 9 k. NH

© Carson-Dellosa CD-3754

Name _____ Skill: Division

Directions
Read each question and choose the correct answer. Mark the space for the answer you have chosen. Mark NH if the answer is not here.

1.

2⟌42

   a. 81
   b. 21
   c. 22
   d. 20
   e. NH

2.

3⟌90

   f. 11
   g. 21
   h. 30
   j. 31
   k. NH

3.

2⟌46

   a. 23
   b. 13
   c. 33
   d. 24
   e. NH

4.

3⟌96

   f. 12
   g. 32
   h. 22
   j. 33
   k. NH

5.

3⟌66

   a. 11
   b. 33
   c. 32
   d. 22
   e. NH

6.

5⟌55

   f. 10
   g. 11
   h. 12
   j. 13
   k. NH

7.

4⟌48

   a. 24
   b. 14
   c. 13
   d. 12
   e. NH

8.

7⟌77

   f. 11
   g. 10
   h. 17
   j. 101
   k. NH

9.

4⟌84

   a. 41
   b. 42
   c. 21
   d. 22
   e. NH

10.

6⟌186

   f. 21
   g. 32
   h. 41
   j. 31
   k. NH

11.

7⟌420

   a. 70
   b. 60
   c. 6
   d. 7
   e. NH

12.

4⟌120

   f. 20
   g. 40
   h. 30
   j. 300
   k. NH

© Carson-Dellosa CD-3754

Name _____   Skill: Division

Directions
Read each question and choose the correct answer. Mark the space for the answer you have chosen. Mark NH if the answer is not here.

1.

3)186

    a. 62
    b. 51
    c. 32
    d. 61
    e. NH

2.

8)488

    f. 81
    g. 61
    h. 71
    j. 60
    k. NH

3.

7)630

    a. 70
    b. 900
    c. 80
    d. 90
    e. NH

4.

9)720

    f. 80
    g. 800
    h. 70
    j. 700
    k. NH

5.

5)250

    a. 50
    b. 5
    c. 500
    d. 15
    e. NH

6.

7)287

    f. 40
    g. 51
    h. 37
    j. 41
    k. NH

7.

4)128

    a. 32
    b. 61
    c. 62
    d. 34
    e. NH

8.

2)486

    f. 143
    g. 223
    h. 243
    j. 263
    k. NH

9.

4)168

    a. 41
    b. 42
    c. 31
    d. 32
    e. NH

10.

3)279

    f. 99
    g. 93
    h. 91
    j. 90
    k. NH

11.

6)546

    a. 61
    b. 80
    c. 91
    d. 81
    e. NH

12.

5)355

    f. 70
    g. 75
    h. 81
    j. 71
    k. NH

© Carson-Dellosa CD-3754

Name _____

Skill: Missing Numbers (Multiplication)

Directions
Read each question and choose the correct answer. Mark the space for the answer you have chosen. Mark NH if the answer is not here.

1. What is the missing number?

    12 x ☐ = 72

    a. 5
    b. 6
    c. 7
    d. 8
    e. NH

2. What is the missing number?

    9 x ☐ = 54

    f. 9
    g. 6
    h. 7
    j. 8
    k. NH

3. What is the missing number?

    9 x ☐ = 81

    a. 9
    b. 8
    c. 7
    d. 6
    e. NH

4. What is the missing number?

    8 x ☐ = 48

    f. 6
    g. 7
    h. 8
    j. 9
    k. NH

5. What is the missing number?

    ☐ x 10 = 80

    a. 5
    b. 6
    c. 7
    d. 8
    e. NH

6. What is the missing number?

    ☐ x 11 = 132

    f. 10
    g. 11
    h. 12
    j. 13
    k. NH

7. What is the missing number?

    ☐ x 7 = 63

    a. 7
    b. 8
    c. 9
    d. 6
    e. NH

8. What is the missing number?

    ☐ x 11 = 77

    f. 0
    g. 1
    h. 8
    j. 7
    k. NH

© Carson-Dellosa CD-3754

Name _____   Skill: Missing Numbers (Division)

Directions
Read each question and choose the correct answer. Mark the space for the answer you have chosen. Mark NH if the answer is not here.

1. What is the missing number?

$$144 \div \square = 12$$

a. 13
b. 12
c. 11
d. 10
e. NH

2. What is the missing number?

$$110 \div \square = 10$$

f. 12
g. 11
h. 10
j. 9
k. NH

3. What is the missing number?

$$56 \div \square = 8$$

a. 7
b. 8
c. 9
d. 10
e. NH

4. What is the missing number?

$$150 \div \square = 5$$

f. 30
g. 40
h. 50
j. 60
k. NH

5. What is the missing number?

$$\square \div 7 = 7$$

a. 49
b. 54
c. 56
d. 63
e. NH

6. What is the missing number?

$$\square \div 12 = 3$$

f. 36
g. 48
h. 54
j. 60
k. NH

7. What is the missing number?

$$\square \div 4 = 9$$

a. 32
b. 36
c. 28
d. 34
e. NH

8. What is the missing number?

$$\square \div 7 = 100$$

f. 7
g. 70
h. 700
j. 7,000
k. NH

© Carson-Dellosa CD-3754

Name _____   Skill: Rounding Numbers

Directions
Read each question and choose the correct answer. Mark the space for the answer you have chosen. Mark NH if the answer is not here.

1. What is 756 rounded to the nearest hundred?

   a. 700
   b. 800
   c. 760
   d. 750
   e. NH

2. What is 5,598 rounded to the nearest hundred?

   f. 5,590
   g. 6,000
   h. 5,600
   j. 5,700
   k. NH

3. What is 2,187 rounded to the nearest hundred?

   a. 2,100
   b. 1,000
   c. 2,200
   d. 3,000
   e. NH

4. What is 89,649 rounded to the nearest hundred?

   f. 90,000
   g. 89,000
   h. 89,700
   j. 89,600
   k. NH

5. What is 1,099 rounded to the nearest thousand?

   a. 0
   b. 900
   c. 1,000
   d. 2,000
   e. NH

6. What is 15,602 rounded to the nearest thousand?

   f. 15,600
   g. 15,700
   h. 16,000
   j. 15,000
   k. NH

7. What is 27,501 rounded to the nearest thousand?

   a. 27,500
   b. 27,000
   c. 28,000
   d. 27,900
   e. NH

8. What is 139,072 rounded to the nearest thousand?

   f. 139,000
   g. 140,000
   h. 100,000
   j. 200,000
   k. NH

© Carson-Dellosa CD-3754

Name _____   Skill: Rounding Numbers

Directions
Read each question and choose the correct answer. Mark the space for the answer you have chosen. Mark NH if the answer is not here.

1. What is 32,650 rounded to the nearest ten thousand?

   a. 33,000
   b. 34,000
   c. 30,000
   d. 40,000
   e. NH

2. What is 76,159 rounded to the nearest ten thousand?

   f. 80,000
   g. 75,000
   h. 70,000
   j. 77,000
   k. NH

3. What is 432,996 rounded to the nearest ten thousand?

   a. 440,000
   b. 400,000
   c. 450,000
   d. 430,000
   e. NH

4. What is 545,832 rounded to the nearest ten thousand?

   f. 540,000
   g. 550,000
   h. 546,000
   j. 500,000
   k. NH

5. What is 929,900 rounded to the nearest hundred thousand?

   a. 1,000,000
   b. 900,000
   c. 900,900
   d. 930,000
   e. NH

6. What is 386,253 rounded to the nearest hundred thousand?

   f. 386,000
   g. 390,000
   h. 300,000
   j. 400,000
   k. NH

7. What is 643,870 rounded to the nearest hundred thousand?

   a. 600,000
   b. 700,000
   c. 650,000
   d. 640,000
   e. NH

8. What is 12,860,000 rounded to the nearest hundred thousand?

   f. 13,900,000
   g. 13,000,000
   h. 12,900,000
   j. 12,850,000
   k. NH

© Carson-Dellosa CD-3754

Name _____   Skill: Decimals/Fractions

Directions
Read each question and choose the correct answer. Mark the space for the answer you have chosen. Mark NH if the answer is not here.

1. What is $\frac{1}{2}$ written as a decimal?

   a. 0.5
   b. 0.12
   c. 1.2
   d. 1.1
   e. NH

5. What is 0.35 written as a fraction?

   a. $\frac{35}{10}$ b. $\frac{35}{1,000}$ c. $\frac{3}{5}$ d. $\frac{35}{100}$ e. NH

2. What is $\frac{3}{4}$ written as a decimal?

   f. 0.34
   g. 3.4
   h. 0.75
   j. 7.5
   k. NH

6. What is 0.052 written as a fraction?

   f. $\frac{5}{2}$ g. $\frac{52}{1,000}$ h. $\frac{52}{10}$ j. $\frac{52}{100}$ k. NH

3. What is $\frac{1}{20}$ written as a decimal?

   a. 0.12
   b. 0.2
   c. 0.05
   d. 0.5
   e. NH

7. What is 0.25 written as a fraction?

   a. $\frac{1}{5}$ b. $\frac{1}{8}$ c. $\frac{4}{25}$ d. $\frac{1}{4}$ e. NH

4. What is $\frac{1}{10}$ written as a decimal?

   f. 1.0
   g. 1.10
   h. 0.1
   j. 0.11
   k. NH

8. What is 0.20 written as a fraction?

   f. $\frac{1}{5}$ g. $\frac{1}{20}$ h. $\frac{1}{2}$ j. $\frac{2}{5}$ k. NH

© Carson-Dellosa CD-3754

Name _____ Skill: Decimals/Fractions

Directions
Read each question and choose the correct answer. Mark the space for the answer you have chosen. Mark NH if the answer is not here.

1. What is $4\frac{1}{2}$ written as a decimal? a. .045 b. 0.45 c. 4.5 d. 45.0 e. NH	5. What is 79.4 written as a fraction? a. $79\frac{1}{4}$ d. $79\frac{4}{10}$ b. $79\frac{4}{100}$ e. NH c. $7\frac{94}{10}$
2. What is $27\frac{3}{10}$ written as a decimal? f. 0.273 g. 27.3 h. 2.73 j. 0.0273 k. NH	6. What is 35.33 written as a fraction? f. $35\frac{3}{10}$ j. $35\frac{3}{3}$ g. $35\frac{33}{10}$ k. NH h. $35\frac{1}{3}$
3. What is $156\frac{21}{100}$ written as a decimal? a. 1.5621 b. 15.621 c. 156.21 d. 1,562.1 e. NH	7. What is 6.04 written as a fraction? a. $6\frac{4}{100}$ d. $6\frac{25}{100}$ b. $6\frac{1}{25}$ e. NH c. $6\frac{4}{10}$
4. What is $27\frac{3}{4}$ written as a decimal? f. 27.75 g. 2.775 h. 0.2775 j. 0.02775 k. NH	8. What is 9.93 written as a fraction? f. $9\frac{93}{1,000}$ j. $9\frac{93}{100}$ g. $9\frac{9}{3}$ k. NH h. $9\frac{9}{10}$

© Carson-Dellosa CD-3754

Name _____

Skill: Computation Practice

Directions
Read each question and choose the correct answer. Mark the space for the answer you have chosen. Mark NH if the answer is not here.

1. 46 +33	a. 78 b. 79 c. 13 d. 73 e. NH	5. 36 12 +21	a. 68 b. 59 c. 58 d. 69 e. NH	9. 3)123	a. 41 b. 40 c. 39 d. 38 e. NH
2. 32 +23	f. 55 g. 51 h. 15 j. 54 k. NH	6. 5 x3	f. 18 g. 17 h. 15 j. 12 k. NH	10. 4)88	f. 23 g. 22 h. 24 j. 25 k. NH
3. 153 +623	a. 730 b. 766 c. 776 d. 736 e. NH	7. 29 x 6	a. 124 b. 174 c. 164 d. 154 e. NH	11. 98 −38	a. 56 b. 68 c. 50 d. 60 e. NH
4. 638 +131	f. 709 g. 569 h. 759 j. 769 k. NH	8. 5)40	f. 9 g. 8 h. 7 j. 6 k. NH	12. 80 −59	f. 31 g. 39 h. 21 j. 29 k. NH

© Carson-Dellosa CD-3754

Name _____   Skill: Computation Practice

Directions
Read each question and choose the correct answer. Mark the space for the answer you have chosen. Mark NH if the answer is not here.

1. 446 − 122	a. 322 b. 324 c. 234 d. 224 e. NH	5. 74 × 2	a. 146 b. 148 c. 158 d. 152 e. NH	9. 900 − 364	a. 536 b. 664 c. 646 d. 546 e. NH
2. 848 − 137	f. 611 g. 715 h. 615 j. 711 k. NH	6. 8)56	f. 3 g. 4 h. 5 j. 6 k. NH	10. 81 × 6	f. 426 g. 421 h. 481 j. 486 k. NH
3. 8 × 4	a. 24 b. 28 c. 32 d. 36 e. NH	7. 9)180	a. 20 b. 200 c. 2 d. 2,000 e. NH	11. 3,674 + 5,108	a. 8,574 b. 8,782 c. 8,772 d. 8,873 e. NH
4. 56 × 3	f. 168 g. 154 h. 158 j. 178 k. NH	8. 4)80	f. 23 g. 22 h. 21 j. 20 k. NH	12. 2)90	f. 41 g. 42 h. 45 j. 43 k. NH

© Carson-Dellosa CD-3754

Name _____   Skill: Computation Practice

Directions
Read each question and choose the correct answer. Mark the space for the answer you have chosen. Mark NH if the answer is not here.

1. What is 2,849 rounded to the nearest thousand?

   a. 2,700
   b. 2,800
   c. 2,000
   d. 3,000
   e. NH

2. What is 214.6 written as a fraction?

   f. $214\frac{6}{100}$ j. $214\frac{46}{10}$
   g. $214\frac{3}{10}$ k. NH
   h. $214\frac{3}{5}$

3. What is the missing number?

   $5 \times \square = 125$

   a. 50
   b. 25
   c. 9
   d. 20
   e. NH

4. What is 156,249 rounded to the nearest hundred?

   f. 156,000
   g. 157,000
   h. 156,300
   j. 156,250
   k. NH

5. What is $9\frac{38}{100}$ written as a decimal?

   a. 938.0
   b. 93.8
   c. 9.38
   d. 0.938
   e. NH

6. What is the missing number?

   $\square \div 7 = 30$

   f. 210
   g. 280
   h. 21
   j. 240
   k. NH

7. What is 724,501 rounded to the nearest thousand?

   a. 723,000
   b. 724,000
   c. 725,000
   d. 724,500
   e. NH

8. What is $47\frac{3}{100}$ written as a decimal?

   f. 0.4703
   g. 4.703
   h. 47.3
   j. 47.03
   k. NH

© Carson-Dellosa CD-3754

Name _____   Skill: Computation Practice

Directions
Read each question and choose the correct answer. Mark the space for the answer you have chosen. Mark NH if the answer is not here.

1. 29 + 30 a. 50 b. 59 c. 19 d. 10 e. NH	5. 22 24 + 13 a. 47 b. 49 c. 59 d. 57 e. NH	9. 640 − 159 a. 519 b. 481 c. 581 d. 419 e. NH
2. 98 + 45 f. 133 g. 143 h. 153 j. 123 k. NH	6. 83 − 51 f. 22 g. 24 h. 34 j. 32 k. NH	10. 5,326 + 1,361 f. 6,645 g. 6,687 h. 7,687 j. 6,085 k. NH
3. 391 + 504 a. 893 b. 895 c. 993 d. 995 e. NH	7. 83 − 56 a. 27 b. 26 c. 37 d. 36 e. NH	11. 1,683 + 659 a. 1,232 b. 2,232 c. 2,332 d. 2,342 e. NH
4. 408 + 139 f. 537 g. 557 h. 647 j. 547 k. NH	8. 896 − 131 f. 775 g. 765 h. 675 j. 767 k. NH	12. 26 31 + 12 f. 59 g. 57 h. 69 j. 67 k. NH

© Carson-Dellosa CD-3754

Name _____   Skill: Computation Practice

Directions
Read each question and choose the correct answer. Mark the space for the answer you have chosen. Mark NH if the answer is not here.

1. 6)240	a. 30 b. 120 c. 40 d. 3 e. NH	5. 7 × 9	a. 63 b. 48 c. 54 d. 56 e. NH	9. 2)74	a. 35 b. 38 c. 36 d. 37 e. NH
2. 4)420	f. 501 g. 105 h. 150 j. 15 k. NH	6. 7 × 4	f. 14 g. 24 h. 28 j. 32 k. NH	10. 4,630 − 1,296	f. 2,334 g. 2,336 h. 3,336 j. 3,334 k. NH
3. 7)567	a. 71 b. 81 c. 78 d. 80 e. NH	7. 64 × 3	a. 182 b. 187 c. 197 d. 192 e. NH	11. 95 × 7	a. 665 b. 635 c. 655 d. 585 e. NH
4. 63 × 6	f. 368 g. 375 h. 378 j. 258 k. NH	8. 11 × 11	f. 1,111 g. 121 h. 112 j. 211 k. NH	12. 9)729	f. 80 g. 81 h. 71 j. 61 k. NH

© Carson-Dellosa CD-3754

Name _____   Skill: Computation Practice

Directions
Read each question and choose the correct answer. Mark the space for the answer you have chosen. Mark NH if the answer is not here.

1. What is 4,709 rounded to the nearest hundred? a. 4,700 b. 4,800 c. 4,710 d. 5,000 e. NH	**5.** What is the missing number? $72 \div \square = 9$ a. 10 b. 9 c. 8 d. 7 e. NH
2. What is the missing number? $\square - 33 = 13$ f. 46 g. 45 h. 40 j. 36 k. NH	**6.** What is 309,999 rounded to the nearest hundred thousand? f. 400,000 g. 300,000 h. 310,000 j. 301,000 k. NH
3. What is $9\frac{4}{10}$ written as a decimal? a. .94 b. 94.0 c. .094 d. 9.4 e. NH	**7.** What is 175.75 written as a fraction? a. $175\frac{1}{3}$ d. $175\frac{3}{4}$ b. $175\frac{1}{4}$ e. NH c. $175\frac{1}{2}$
4. What is 5,499 rounded to the nearest hundred? f. 5,000 g. 5,500 h. 4,990 j. 5,400 k. NH	**8.** What is 4,179,821 rounded to the nearest hundred thousand? f. 4,000,000 g. 4,179,800 h. 4,179,900 j. 4,180,000 k. NH

Name _____   Skill: Word Problems

Directions
Read each question and choose the correct answer. Mark the space for the answer you have chosen. Mark NH if the answer is not here.

1. A touchdown in football is worth 6 points. What number sentence tells how many points would be earned for 5 touchdowns?
   a. 6 + 5 = n
   b. 6 − 5 = n
   c. 6 ÷ 5 = n
   d. 6 x 5 = n
   e. NH

2. Kelly types 30 words a minute. What number sentence tells how many words she can type in 20 minutes?
   f. 30 + 20 = n
   g. 30 − 20 = n
   h. 30 ÷ 20 = n
   j. 30 x 20 = n
   k. NH

3. Fran needed 42 hangers. She had 23. What number sentence tells how many more hangers she needs?
   a. 42 + 23 = n
   b. 42 x 23 = n
   c. 42 ÷ 23 = n
   d. 42 − 23 = n
   e. NH

4. Jude has 72 plants. He wants to put 8 in each row. What number sentence tells how many rows he'll need?
   f. 72 + 8 = n
   g. 72 x 8 = n
   h. 72 ÷ 8 = n
   j. 72 − 8 = n
   k. NH

5. Bill must put 5 dimes in each slot. He has 20 slots. What sentence tells how many dimes he needs?
   a. 20 + 5 = n
   b. 20 x 5 = n
   c. 20 ÷ 5 = n
   d. 20 − 5 = n
   e. NH

6. I have 32 pennies, 43 dimes, and 65 nickels. What number sentence tells how many coins I have?
   f. 65 + 43 + 32 = n
   g. 65 x 43 x 32 = n
   h. (65 ÷ 32) + 43 = n
   j. (65 − 32) − 43 = n
   k. NH

7. Jack gave away 76 of his cards. He now has 112 left. What number sentence tells how many cards he had at first?
   a. 112 + 76 = n
   b. 112 x 76 = n
   c. 112 ÷ 76 = n
   d. 112 − 76 = n
   e. NH

8. Ben gave each of his 12 friends 3 balls. What sentence tells how many balls Ben gave away?
   f. 12 + 3 = n
   g. 12 x 3 = n
   h. 12 ÷ 3 = n
   j. 12 − 3 = n
   k. NH

© Carson-Dellosa CD-3754

Name _____

Skill: Word Problems

Directions
Read each question and choose the correct answer. Mark the space for the answer you have chosen. Mark NH if the answer is not here.

---

1. Mark bought 7 pencils at 25¢ each. There was no tax. How much did the seven pencils cost?

a. $2.00
b. $1.75
c. $1.50
d. $1.25
e. NH

5. Perry bought $\frac{1}{3}$ of a pound of candy. It cost him $4.50. At that rate, how much does one pound of candy cost?

a. $1.50
b. $3.00
c. $13.50
d. $18.00
e. NH

---

2. Myron paid $5.50 for five pads of construction paper. How much did each pad cost?

f. $1.10
g. 55¢
h. $1.01
j. $2.75
k. NH

6. Germaine bought 5 CDs for $45.00. At that rate, what was the cost for each CD?

f. $6.00
g. $7.00
h. $8.00
j. $9.00
k. NH

---

3. Sylvia has 3 five dollar bills, 5 quarters, 15 dimes, and 3 nickels. How much money does Sylvia have?

a. $26.00
b. $17.90
c. $15.80
d. $16.10
e. NH

7. Paulie donated $12 to each of his 5 favorite charities. How much money did Paulie donate in all?

a. $2.40
b. $48.00
c. $60.00
d. $72.00
e. NH

---

4. The blue sweater costs $24.36. The red one costs $35.80. How much more does the red sweater cost?

f. $60.16
g. $11.56
h. $12.56
j. $11.44
k. NH

8. Karin had $315.25 in her savings. She deposited $75.45 today. What is the new balance in her savings?

f. $239.80
g. $390.70
h. $380.65
j. $380.25
k. NH

© Carson-Dellosa CD-3754

Name _____   Skill: Word Problems

Directions
Read each question and choose the correct answer. Mark the space for the answer you have chosen. Mark NH if the answer is not here.

1. There were 97 birds on the lawn. 53 flew into the trees. How many birds were left on the lawn?

   a. 150 birds
   b. 54 birds
   c. 50 birds
   d. 44 birds
   e. NH

2. Max ran 10 miles on Monday, 12 on Tuesday and 14 on Wednesday. How many miles did Max run in all?

   f. 22 miles
   g. 24 miles
   h. 36 miles
   j. 39 miles
   k. NH

3. Joan had 723 bottle caps. She gave some away and has 684 left. How many did Joan give away?

   a. 61 bottle caps
   b. 49 bottle caps
   c. 39 bottle caps
   d. 21 bottle caps
   e. NH

4. Nan put 12 chairs in each row. There were 24 rows. How many chairs did Nan use?

   f. 240 chairs
   g. 360 chairs
   h. 288 chairs
   j. 260 chairs
   k. NH

5. Macy had to sort 161 candies evenly into 7 jars. How many candies did she put in each jar?

   a. 23 candies
   b. 22 candies
   c. 21 candies
   d. 20 candies
   e. NH

6. Moe has to read a book with 964 pages. He has finished 432. How many pages are left for him to read?

   f. 596 pages
   g. 592 pages
   h. 532 pages
   j. 492 pages
   k. NH

7. Robyn planted 12 rows of trees. She had a total of 312 trees. How many trees were in each row?

   a. 324 trees
   b. 26 trees
   c. 300 trees
   d. 25 trees
   e. NH

8. 104 people voted "yes." 213 people voted "no." How many people voted?

   f. 315 people
   g. 111 people
   h. 311 people
   j. 317 people
   k. NH

© Carson-Dellosa CD-3754

Name _____   Skill: Word Problems

Directions
Read each question and choose the correct answer. Mark the space for the answer you have chosen. Mark NH if the answer is not here.

---

1. We had 147 marbles. 97 of them were blue and the rest were green. How many green marbles are there?
   a. 50 marbles
   b. 54 marbles
   c. 57 marbles
   d. 53 marbles
   e. NH

2. Mom baked 144 rolls with 12 on each pan. How many pans of rolls did Mom bake?
   f. 10 pans
   g. 12 pans
   h. 72 pans
   j. 132 pans
   k. NH

3. Iris drew 3 pictures for each of her 6 uncles. How many pictures did Iris draw?
   a. 3 pictures
   b. 6 pictures
   c. 12 pictures
   d. 18 pictures
   e. NH

4. Jill ate 1,256 calories today and 2,453 calories yesterday. How many calories did she eat both days?
   f. 3,803 calories
   g. 3,709 calories
   h. 2,709 calories
   j. 1,197 calories
   k. NH

5. Phyllis grew 123 roses and 207 tulips this year. How many flowers did she grow in all?
   a. 84 roses and tulips
   b. 324 roses and tulips
   c. 330 roses and tulips
   d. 344 roses and tulips
   e. NH

6. It is 157 miles to Mahew and 123 miles to DeVille. How much further is Mahew than DeVille?
   f. 30 miles
   g. 34 miles
   h. 74 miles
   j. 280 miles
   k. NH

7. Dan scored 7 goals today. If each goal is worth 5 points, how many points did Dan score?
   a. 35 points
   b. 30 points
   c. 12 points
   d. 2 points
   e. NH

8. Winston collected 172 pounds of paper and 36 pounds of metal. How much more paper than metal has he?
   f. 208 pounds
   g. 174 pounds
   h. 144 pounds
   j. 136 pounds
   k. NH

© Carson-Dellosa CD-3754

Name _____

Skill: Word Problems

Directions
Read each question and choose the correct answer. Mark the space for the answer you have chosen. Mark NH if the answer is not here.

Use the information given here to answer questions 1-3.

Davis and Louis went to the fair. Davis spent $17.50 on rides and $22.00 on games. He won 6 prizes. Louis spent $45.00. They were there from noon until 6:00 p.m.

1. How much more money did Louis spend than Davis?

   a. $27.50
   b. $23.00
   c. $5.50
   d. $4.50
   e. NH

2. What piece of information is <u>not</u> necessary to find out how much money both boys spent?

   f. Davis spent $17.50 on rides
   g. Davis spent $22.00 on games
   h. Davis won 6 prizes
   j. Louis spent $45.00
   k. NH

3. How long were the boys at the fair?

   a. 7 hours
   b. 6 hours
   c. 5 hours
   d. 4 hours
   e. NH

Use the information given here to answer questions 4-6.

Kaitlin and Evelyn gave a party. They bought 240 sandwiches and 90 sodas. They invited 35 guests, but only 20 came.

4. How many sandwiches could each of the 20 guests have if they were split evenly?

   f. 12 sandwiches
   g. 3 sandwiches
   h. 95 sandwiches
   j. 220 sandwiches
   k. NH

5. What information is <u>not</u> necessary to find out how many sandwiches each guest could have?

   a. 240 sandwiches
   b. 90 sodas
   c. 20 guests came
   d. each guest got the same amount
   e. NH

6. How many guests were invited but did not come to the party?

   f. 70 guests did not come
   g. 55 guests did not come
   h. 15 guests did not come
   j. 12 guests did not come
   k. NH

© Carson-Dellosa CD-3754

Name _____

Skill: Shapes

Directions
Read each question and choose the correct answer. Mark the space for the answer you have chosen. Mark NH if the answer is not here.

1. What shape is symmetrical ?

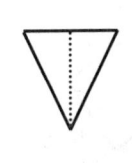

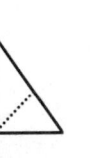

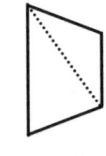

a.   b.   c.   d.   NH e.

2. What shape has four equal sides?

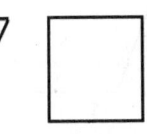

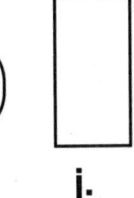

f.   g.   h.   j.   NH k.

3. What shape has four equal angles?

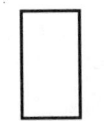

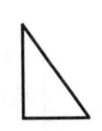

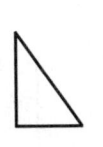

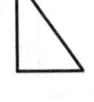

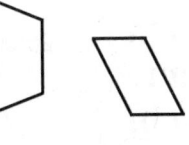

a.   b.   c.   d.   NH e.

4. What shape is symmetrical?

f.   g.   h.   j.   NH k.

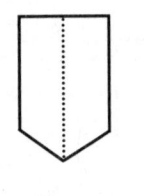

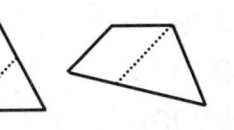

5. What shape is <u>not</u> a square?

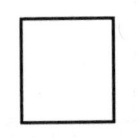

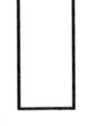

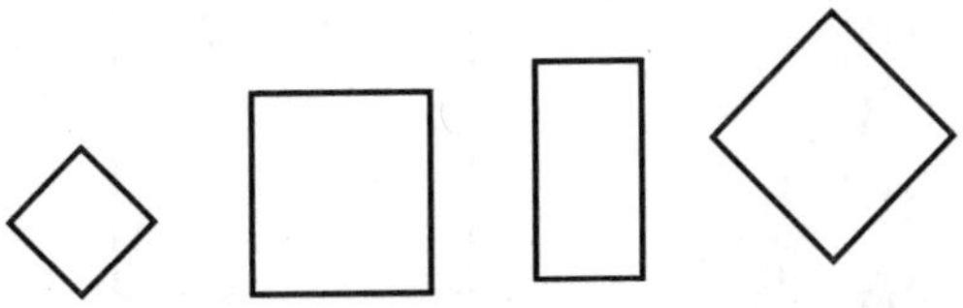

a.   b.   c.   d.   NH e.

6. What figure is a rectangular prism?

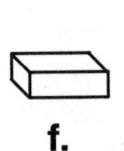

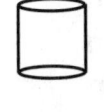

f.   g.   h.   j.   NH k.

7. What figure is a cone?

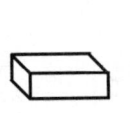

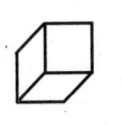

a.   b.   c.   d.   NH e.

8. What figure is a cube?

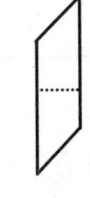

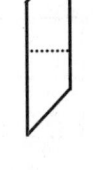

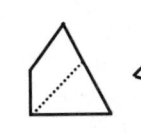

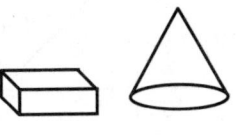

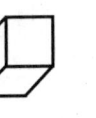

f.   g.   h.   j.   NH k.

© Carson-Dellosa CD-3754   72

Name _____   Skill: Perimeter

Directions
Read each question and choose the correct answer. Mark the space for the answer you have chosen. Mark NH if the answer is not here.

1. What is the perimeter of this square?

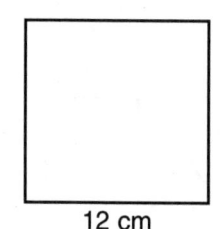

a. 12 cm
b. 24 cm
c. 36 cm
d. 48 cm
e. NH

12 cm

5. What is the perimeter of this rectangle?

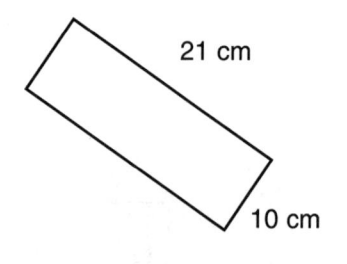

a. 31 cm
b. 62 cm
c. 210 cm
d. 42 cm
e. NH

21 cm
10 cm

2. What is the perimeter of this rectangle?

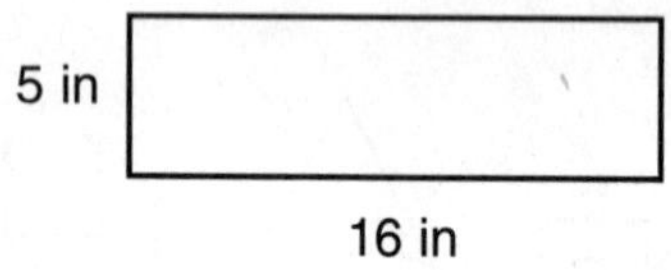

f. 80 in
g. 42 in
h. 32 in
j. 21 in
k. NH

5 in
16 in

6. What is the perimeter of this hexagon?

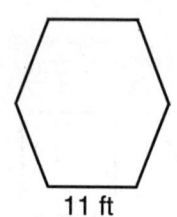

f. 44 ft
g. 88 ft
h. 55 ft
j. 77 ft
k. NH

11 ft

3. What is the perimeter of this square?

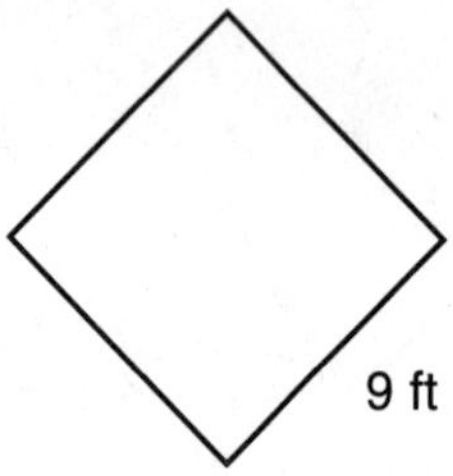

a. 9 feet
b. 18 feet
c. 45 feet
d. 54 feet
e. NH

9 ft

7. What is the perimeter of this shape?

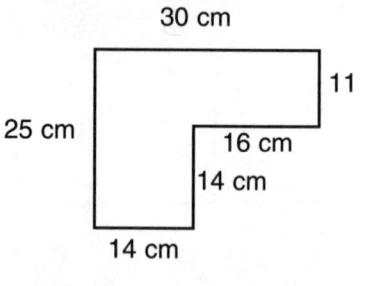

a. 86 cm
b. 110 cm
c. 96 cm
d. 112 cm
e. NH

30 cm
11
25 cm
16 cm
14 cm
14 cm

4. What is the perimeter of this shape?

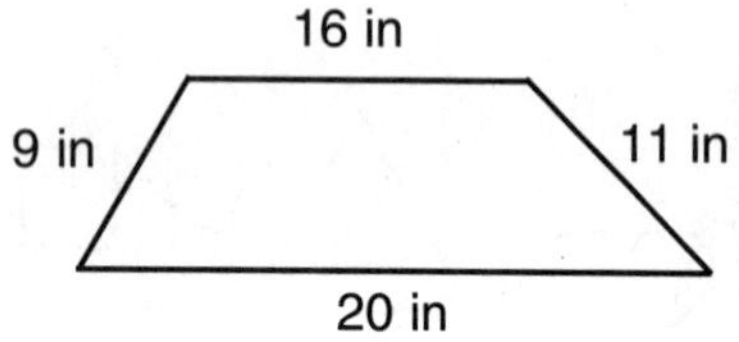

f. 56 in
g. 47 in
h. 40 in
j. 20 in
k. NH

16 in
9 in   11 in
20 in

8. What is the perimeter of this square?

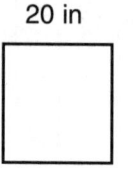

f. 40 in
g. 60 in
h. 80 in
j. 100 in
k. NH

20 in

© Carson-Dellosa CD-3754   73

Name _____ Skill: Area

Directions
Read each question and choose the correct answer. Mark the space for the answer you have chosen. Mark NH if the answer is not here.

1. What is the area of this square?

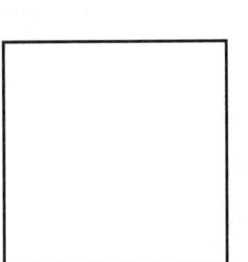

a. 15 sq in
b. 60 sq in
c. 150 sq in
d. 225 sq in
e. NH

15 in

2. What is the area of this rectangle?

2 ft

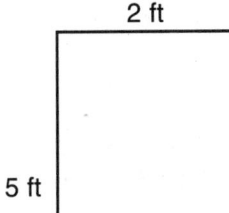

f. 7 sq ft
g. 14 sq ft
h. 29 sq ft
j. 10 sq ft
k. NH

5 ft

3. What is the area of this rectangle?

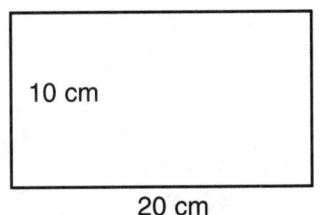

a. 100 sq cm
b. 200 sq cm
c. 30 sq cm
d. 60 sq cm
e. NH

10 cm
20 cm

4. What is the area of this square?

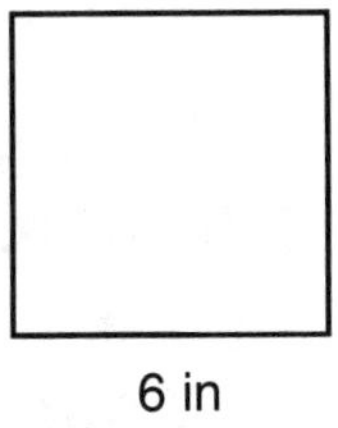

f. 6 sq in
g. 12 sq in
h. 34 sq in
j. 36 sq in
k. NH

6 in

5. What is the area of this rectangle?

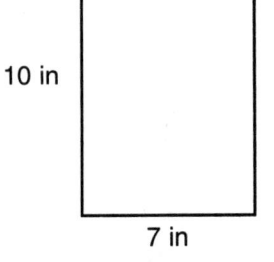

a. 17 sq in
b. 34 sq in
c. 70 sq in
d. 77 sq in
e. NH

10 in
7 in

6. What is the area of this square?

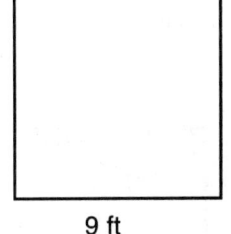

f. 100 sq ft
g. 90 sq ft
h. 36 sq ft
j. 18 sq ft
k. NH

9 ft

7. What is the area of this rectangle?

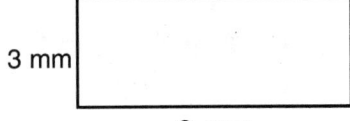

a. 11 sq mm
b. 22 sq mm
c. 24 sq mm
d. 32 sq mm
e. NH

3 mm
8 mm

8. What is the area of this shape?

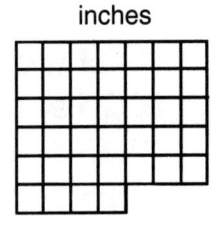

inches

f. 42 sq in
g. 39 sq in
h. 36 sq in
j. 33 sq in
k. NH

© Carson-Dellosa CD-3754

Name _____ Skill: Time

Directions
Read each question and choose the correct answer. Mark the space for the answer you have chosen. Mark NH if the answer is not here.

---

1. Howie cooks the cake for 25 minutes. He put it in the oven at 1:30. At what time will he take it out?

 a. 2:00
 b. 1:45
 c. 1:55
 d. 2:05
 e. NH

5. Look at the clock. It is p.m. What time will it be in 6 hours?

 a. 8:00 p.m.
 b. 9:30 p.m.
 c. 9:00 p.m.
 d. 8:30 p.m.
 e. NH

---

2. Janice ran for 50 minutes. She ended at 9:00 a.m. At what time did she start running?

 f. 7:50 a.m.
 g. 8:10 a.m.
 h. 9:50 a.m.
 j. 7:10 a.m.
 k. NH

6. Look at the clock. It is a.m. What time will it be in 6 hours 15 minutes?

 f. 2:20 p.m.
 g. 8:15 a.m.
 h. 2:10 p.m.
 j. 3:20 p.m.
 k. NH

---

3. Pete started his homework at 3:35 and finished at 4:15. How long did his homework take?

 a. 1 hour and 20 minutes
 b. 45 minutes
 c. 20 minutes
 d. 30 minutes
 e. NH

7. Look at the clock. It is a.m. What time will it be in 3 hours and 35 minutes?

 a. 9:35 a.m.
 b. 10:25 p.m.
 c. 2:15 p.m.
 d. 10:35 a.m.
 e. NH

---

4. Carol must read for 45 minutes each day. If she begins at 11:40 a.m., at what time can she stop?

 f. 12:30 a.m.
 g. 12:25 a.m.
 h. 12:30 p.m.
 j. 12:25 p.m.
 k. NH

8. Look at the clock. It is a.m. What time was it 3 hours and 10 minutes ago?

 f. 6:45 a.m.
 g. 1:50 a.m.
 h. 1:45 a.m.
 j. 8:40 a.m.
 k. NH

© Carson-Dellosa CD-3754

Name _____   Skill: Time

Directions
Read each question and choose the correct answer. Mark the space for the answer you have chosen. Mark NH if the answer is not here.

1. The minute hand is not shown on this clock. If the time is 5:15, at what number should the minute hand point?
   a. 2
   b. 3
   c. 6
   d. 9
   e. NH

5. The hour hand is not shown on this clock. If the time is 4:25, where should the hour hand point?
   a. between 3 and 4
   b. just before the 6
   c. between 4 and 5
   d. just after the 3
   e. NH

2. The minute hand is not shown on this clock. If the time is 3:45, at what number should the minute hand point?
   f. 9
   g. 8
   h. 4
   j. 3
   k. NH

6. The hour hand is not shown on this clock. If the time is 7:55, where should the hour hand point?
   f. just before the 7
   g. between 8 and 9
   h. just before the 8
   j. to the 7
   k. NH

3. The minute hand is not shown on this clock. If the time is 2:10, at what number should the minute hand point?
   a. 1
   b. 2
   c. 10
   d. 11
   e. NH

7. The hour hand is not shown on this clock. If the time is 12:30, where should the hour hand point?
   a. to the 12
   b. to the 1
   c. between 12 and 1
   d. just after the 1
   e. NH

4. The minute hand is not shown on this clock. If the time is 10:35, at what number should the minute hand point?
   f. 4
   g. 5
   h. 6
   j. 7
   k. NH

8. The hour hand is not shown on this clock. If the time is 9:25, where should the hour hand point?
   f. just before the 9
   g. just before the 1
   h. between 9 and 10
   j. just after the 10
   k. NH

© Carson-Dellosa CD-3754

Name _____ Skill: Money

Directions
Read each question and choose the correct answer. Mark the space for the answer you have chosen. Mark NH if the answer is not here.

1. How much money is shown?

a. 59¢
b. 85¢
c. 89¢
d. $1.04
e. NH

5. How much money is shown?

a. $3.02
b. $3.15
c. $3.35
d. $3.50
e. NH

2. How much money is shown?

f. $1.29
g. $1.33
h. $1.45
j. $1.60
k. NH

6. How much money is shown?

f. $2.65
g. $6.65
h. $6.61
j. $6.56
k. NH

3. How much money is shown?

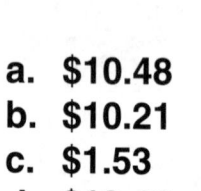

a. 88¢
b. $1.54
c. $1.64
d. $1.79
e. NH

7. How much money is shown?

a. $10.48
b. $10.21
c. $1.53
d. $10.43
e. NH

4. How much money is shown?

f. 73¢
g. $1.08
h. $1.13
j. $1.45
k. NH

8. How much money is shown?

f. $3.25
g. $16.25
h. $15.25
j. $30.25
k. NH

© Carson-Dellosa CD-3754

Name _____   Skill: Money

Directions
Read each question and choose the correct answer. Mark the space for the answer you have chosen. Mark NH if the answer is not here.

---

1. Tony's purchase cost $7.56. How much change will he get from a ten-dollar bill?

   a. $2.54
   b. $2.44
   c. $3.54
   d. $3.44
   e. NH

5. Justin's purchase cost $16.87. How much change will he get from a twenty-dollar bill?

   a. $3.13
   b. $3.23
   c. $4.13
   d. $4.23
   e. NH

2. Hannah's purchase cost $3.27. How much change will she get from a five-dollar bill?

   f. $1.73
   g. $1.83
   h. $1.82
   j. $2.83
   k. NH

6. Meghan's purchase cost $12.18. How much change will she get from a twenty-dollar bill?

   f. $8.92
   g. $8.82
   h. $7.82
   j. $7.92
   k. NH

3. Ralph's purchase cost $4.39. How much change will he get from a ten-dollar bill?

   a. $4.71
   b. $6.71
   c. $5.61
   d. $5.71
   e. NH

7. Andrew's purchase cost $4.76. How much change will he get from a twenty-dollar bill?

   a. $15.34
   b. $15.24
   c. $16.24
   d. $16.34
   e. NH

4. Fran's purchase cost $11.05. How much change will she get from a twenty-dollar bill?

   f. $9.95
   g. $9.85
   h. $8.85
   j. $8.95
   k. NH

8. Vicki's purchase cost $13.51. How much change will she get from a twenty-dollar bill?

   f. $7.59
   g. $7.49
   h. $6.59
   j. $6.49
   k. NH

© Carson-Dellosa CD-3754

Name _____   Skill: Metric Conversion

Directions
Read each question and choose the correct answer. Mark the space for the answer you have chosen. Mark NH if the answer is not here.

1. 1 meter = a. 1 millimeter b. 10 millimeters c. 100 millimeter d. 1,000 millimeters e. NH	**5.** 10 millimeters = a. 0.1 centimeters b. 1 centimeter c. 10 centimeters d. 100 centimeters e. NH
2. 1,000 grams = f. 1,000 kilograms g. 10 kilograms h. 1 kilogram j. 0.1 kilogram k. NH	**6.** 100 centigrams = f. 0.1 gram g. 1 gram h. 10 grams j. 100 grams k. NH
3. 10 centimeters = a. 1,000 millimeters b. 1 millimeter c. 10 millimeters d. 100 millimeters e. NH	**7.** 1,000 milliliters = a. 0.1 liter b. 100 liters c. 10 liters d. 1 liter e. NH
4. 1 kilometer = f. 10,000 meters g. 1,000 meters h. 100 meters j. 10 meters k. NH	**8.** 100 centimeters = f. 0.1 meter g. 1 meter h. 10 meters j. 100 meters k. NH

© Carson Dellosa CD-3754

Name _____

Skill: Metric Measurement

Directions
Read each question and choose the correct answer. Mark the space for the answer you have chosen. Mark NH if the answer is not here.

1. What metric unit would be used to measure the weight of a cup? a. liter b. millimeter c. gram d. kilogram e. NH	**5.** What metric unit would be used to measure the distance between cities? a. kiloliters b. grams c. kilometers d. centimeter e. NH
2. What metric unit would be used to measure how tall a baby is? f. centigram g. centimeter h. meter j. kiloliter k. NH	**6.** What metric unit would be used to measure the volume of water in a bucket? f. milligrams g. liters h. centimeters j. kilograms k. NH
3. What metric unit would be used to measure the weight of a truck? a. grams b. liters c. kiloliters d. kilograms e. NH	**7.** What metric unit would be used to measure the weight of a calculator? a. grams b. liters c. meters d. centimeters e. NH
4. What metric unit would be used to measure a spoonful of sugar? f. kilograms g. milligrams h. kiloliters j. liters k. NH	**8.** What metric unit would be used to measure the weight of a feather? f. milligrams g. grams h. kilograms j. millimeters k. NH

© Carson-Dellosa CD-3754

Name _____

Skill: Metric Measurement

Directions
Read each question and choose the correct answer. Mark the space for the answer you have chosen. Mark NH if the answer is not here.

---

1. 10 millimeters =

a. 0.001 kilometer
b. 1 centimeter
c. 1,000 centimeters
d. 0.1 meter
e. NH

5. 53,000 millimeters =

a. 530 centimeters
b. 53 meters
c. 5.3 kilometers
d. 53 kilometers
e. NH

---

2. 5 liters =

f. 500 kiloliters
g. 50 milliliters
h. 5 kiloliters
j. 500 centiliters
k. NH

6. 0.07 kilograms =

f. 7 grams
g. 7,000 centigrams
h. 700 milligrams
j. 700 grams
k. NH

---

3. 16 centigrams =

a. 160 milligrams
b. 1,600 milligrams
c. 160 grams
d. 1.6 grams
e. NH

7. 14 centimeters =

a. 14,000 millimeters
b. 140 meters
c. 140 millimeters
d. 1.4 meters
e. NH

---

4. 2,500 meters =

f. 25 centimeters
g. 25 kilometers
h. 2.5 kilometers
j. 250,000 centimeters
k. NH

8. 350 centigrams =

f. 3.5 grams
g. 35 grams
h. 35 kilograms
j. 3.5 kilograms
k. NH

© Carson-Dellosa CD-3754

Name _____   Skill: Bar Graph

Directions
Read each question and choose the correct answer. Mark the space for the answer you have chosen. Mark NH if the answer is not here.

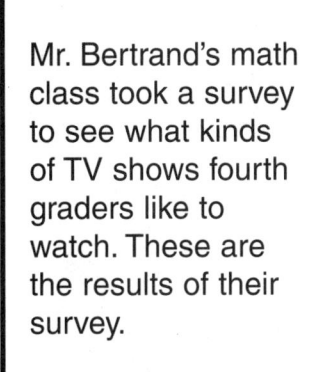

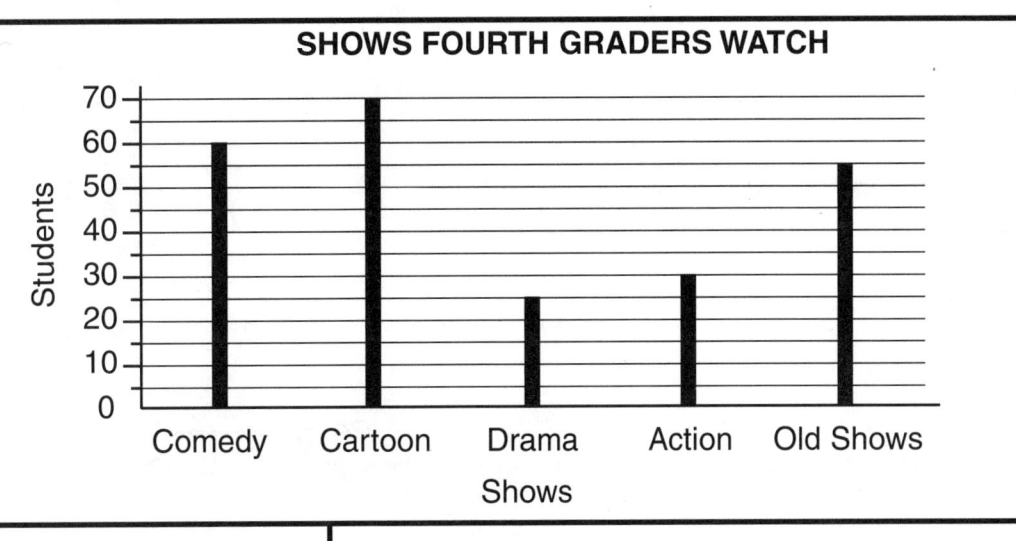

Mr. Bertrand's math class took a survey to see what kinds of TV shows fourth graders like to watch. These are the results of their survey.

SHOWS FOURTH GRADERS WATCH

(Bar graph – Students vs Shows: Comedy 60, Cartoon 70, Drama 25, Action 30, Old Shows 55)

---

1. How many fourth graders participated in this survey?

   a. 230
   b. 235
   c. 240
   d. 245
   e. NH

2. What is the most popular type of show among these fourth graders?

   f. Cartoon
   g. Comedy
   h. Old shows
   j. Drama
   k. NH

3. What type of show got five more votes than drama?

   a. Comedy
   b. Old shows
   c. Cartoon
   d. Action
   e. NH

4. How many students like to watch old shows?

   f. 60
   g. 55
   h. 40
   j. 35
   k. NH

5. How many more students chose comedy than drama?

   a. 25
   b. 30
   c. 35
   d. 40
   e. NH

6. How many students voted for cartoons or action?

   f. 100
   g. 140
   h. 90
   j. 85
   k. NH

© Carson-Dellosa CD-3754

Name _____ Skill: Pie Graph

Directions
Read each question and choose the correct answer. Mark the space for the answer you have chosen. Mark NH if the answer is not here.

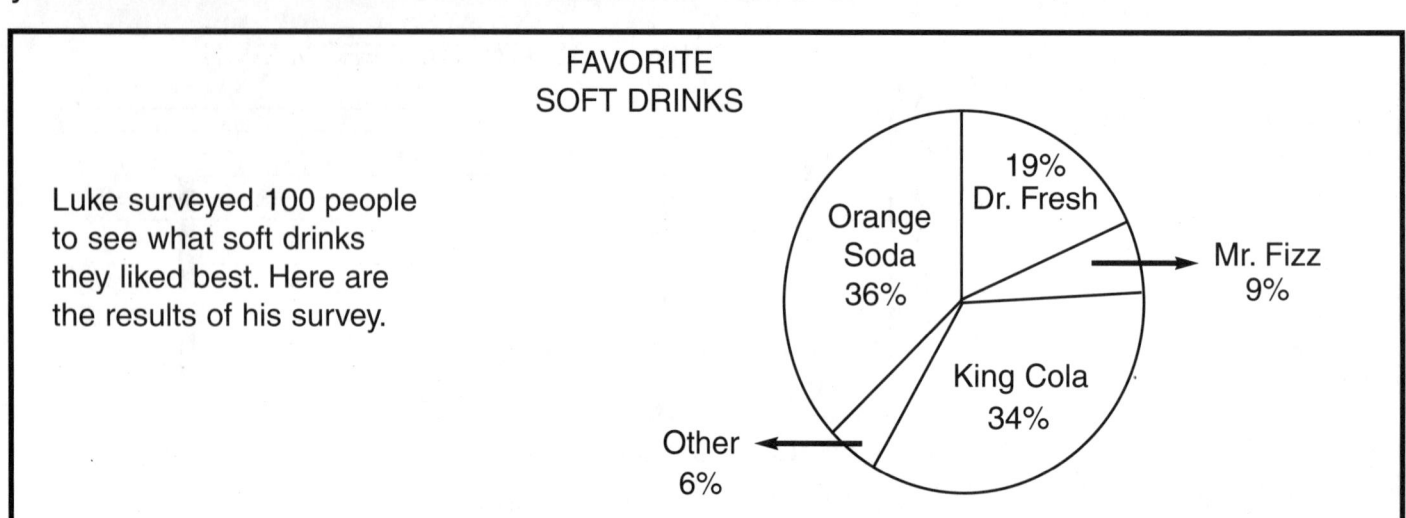

FAVORITE SOFT DRINKS

Luke surveyed 100 people to see what soft drinks they liked best. Here are the results of his survey.

- Orange Soda 36%
- Dr. Fresh 19%
- Mr. Fizz 9%
- King Cola 34%
- Other 6%

1. What soft drink was most popular with the people in this survey?

- a. Orange Soda
- b. King Cola
- c. Dr. Fresh
- d. Mr. Fizz
- e. NH

2. By how much did the leading soft drink beat the second most popular drink?

- f. 17%
- g. 9%
- h. 2%
- j. 1%
- k. NH

3. What percentage of the people liked either Dr. Fresh or Mr. Fizz?

- a. 9%
- b. 15%
- c. 19%
- d. 28%
- e. NH

4. What two groups make up exactly 40% of this graph?

- f. Orange Soda and Other
- g. Orange Soda and Mr. Fizz
- h. Dr. Fresh and Orange Soda
- j. King Cola And Other
- k. NH

5. How many people voted for Dr. Fresh?

- a. 38 people
- b. 19 people
- c. 15 people
- d. 9 people
- e. NH

6. Who were the people that Luke surveyed?

- f. students
- g. business people
- h. people at the mall
- j. teachers
- k. NH

© Carson-Dellosa CD-3754

Name _____   Skill: Pictograph

Directions
Read each question and choose the correct answer. Mark the space for the answer you have chosen. Mark NH if the answer is not here.

Five friends had a contest to see who could earn the most A's during the school year. The results are shown on this chart.

Key: ★ = 5 A's

Students	NUMBER OF A'S EARNED THIS YEAR
Amanda C.	★★★★★★★★★★★★★
Jennie D.	★★★★★★★
Patrick M.	★★★★★★★★
Doneshia R.	★★★★★★★★★★★★
Jordan S.	★★★★★★★★★★★★★

1. Who earned the most A's this year?

   a. Jordan
   b. Jordan and Amanda
   c. Patrick
   d. Doneshia
   e. NH

2. How many A's did Doneshia earn?

   f. 60 A's
   g. 55 A's
   h. 35 A's
   j. 12 A's
   k. NH

3. In what subjects did Amanda get most of her A's?

   a. English
   b. Math
   c. Science
   d. Social Studies
   e. NH

4. How many more A's did Jordan earn than Jennie?

   f. 6 more A's
   g. 15 more A's
   h. 30 more A's
   j. 35 more A's
   k. NH

5. How many A's did Patrick and Doneshia earn in all?

   a. 90 A's
   b. 95 A's
   c. 100 A's
   d. 105 A's
   e. NH

6. How many A's did these five students earn altogether?

   f. 280 A's
   g. 270 A's
   h. 175 A's
   j. 54 A's
   k. NH

Name _____ Skill: Pictographs

Directions
Read each question and choose the correct answer. Mark the space for the answer you have chosen. Mark NH if the answer is not here.

Mr. Martinez wants to know how many band instruments to order for next fall. He asked the band members to take a survey to find out what instruments they will be playing. Here are the results.

Key: ✔ = ten students

Instruments	BAND INSTRUMENTS FOR NEXT YEAR
Trombone	✔✔✔✔✔✔✔✔✔
Flute	✔✔✔✔✔✔✔
Drum	✔✔✔✔✔
Clarinet	✔✔✔✔✔✔✔✔
Saxophone	✔✔✔✔✔✔✔✔✔✔

1. Mr. Martinez will order the greatest number of which instrument?

   a. trombone
   b. flute
   c. clarinet
   d. saxophone
   e. NH

2. How many students said they will be playing the clarinet next fall?

   f. 11 students
   g. 12 students
   h. 110 students
   j. 120 students
   k. NH

3. What instrument will have the least number of students playing it?

   a. flute
   b. drum
   c. clarinet
   d. trombone
   e. NH

4. If everyone in the survey returns to the band next year, how many students will be in the band?

   f. 490 students
   g. 470 students
   h. 94 students
   j. 47 students
   k. NH

5. What instruments were chosen by less than 100 students?

   a. drum and trombone
   b. flute and clarinet
   c. flute and drum
   d. trombone and flute
   e. NH

6. How many more students chose the saxophone than the drum?

   f. 7 students
   g. 14 students
   h. 35 students
   j. 70 students
   k. NH

© Carson-Dellosa CD-3754

Name _____   Skill: Application Practice

Directions
Read each question and choose the correct answer. Mark the space for the answer you have chosen. Mark NH if the answer is not here.

1. Joey paid $29.85 for three books. How much did each book cost?

   a. $8.95
   b. $9.95
   c. $9.45
   d. $7.45
   e. NH

5. A flock of geese migrated 20 miles a day for five days. How far did the geese migrate?

   a. 100 miles
   b. 50 miles
   c. 200 miles
   d. 55 miles
   e. NH

2. How much money is this?

   f. $1.08
   g. $1.18
   h. $1.23
   j. $1.13
   k. NH

6. Sarah is reading a 925-page book. She has 218 pages to go. How many pages has Sarah read?

   f. 707
   g. 777
   h. 897
   j. 717
   k. NH

3. Look at the clock. It is p.m. What time will it be in 2 hours and 45 minutes?

   a. 5:45 p.m.
   b. 2:45 p.m.
   c. 3:00 p.m.
   d. 3:45 p.m.
   e. NH

7. Which shape is symmetrical?

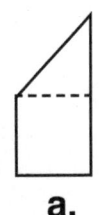

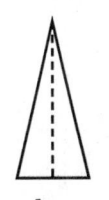

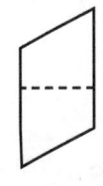

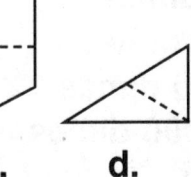

   a.   b.   c.   d.   e. NH

4. What is the area of this rectangle?

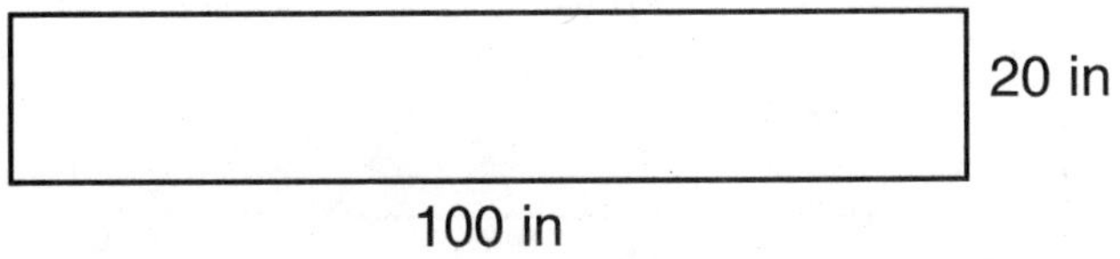

   20 in
   100 in

   f. 200 sq in   j. 2,000 sq in
   g. 240 sq in   k. NH
   h. 220 sq m

8. What is the perimeter of this rectangle?

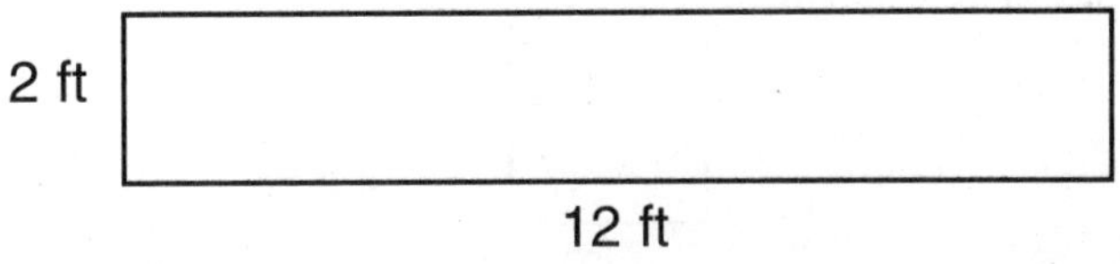

   2 ft
   12 ft

   f. 14 ft   j. 280 ft
   g. 24 ft   k. NH
   h. 28 ft

© Carson-Dellosa CD-3754

Name _____

Skill: Application Practice

Directions
Read each question and choose the correct answer. Mark the space for the answer you have chosen. Mark NH if the answer is not here.

1. Henry watches 30 minutes of TV every day. How much TV does he watch in one week?

    a. 2 hours
    b. 1.5 hours
    c. 3 hours
    d. 3.5 hours
    e. NH

2. Christine types 42 words a minute. How many words can she type in 5 minutes?

    f. 85 words
    g. 210 words
    h. 120 words
    j. 280 words
    k. NH

3. Carl has $20.00 in dimes. How many dimes does he have?

    a. 2 dimes
    b. 20 dimes
    c. 200 dimes
    d. 2,000 dimes
    e. NH

4. What is the area of this square?

    7 m

    f. 14 sq m   j. 28 sq m
    g. 21 sq m   k. NH
    h. 49 sq m

5. Jamie skated 6 hours Friday, 6 hours Saturday, and 7 hours Sunday. How many hours did Jamie skate in all?

    a. 17 hours
    b. 18 hours
    c. 19 hours
    d. 20 hours
    e. NH

6. Norm's purchase cost $18.33. How much change will he get from a twenty-dollar bill?

    f. $1.67
    g. $1.77
    h. $.77
    j. $2.77
    k. NH

7. 5 kilometers =

    a. 5,000 meters
    b. 5,000 millimeters
    c. 5,000 centimeters
    d. 1 liter
    e. NH

8. What is the perimeter of this shape?

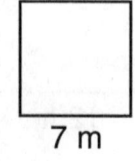

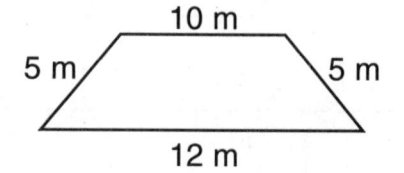

    10 m
    5 m   5 m
    12 m

    f. 14 ft   j. 280 ft
    g. 24 ft   k. NH
    h. 28 ft

© Carson-Dellosa CD-3754

Name _____

Skill: Application Practice

Directions
Read each question and choose the correct answer. Mark the space for the answer you have chosen. Mark NH if the answer is not here.

Christy asked 100 people to pick their favorite color. Here are the results of the survey.

FAVORITE COLORS

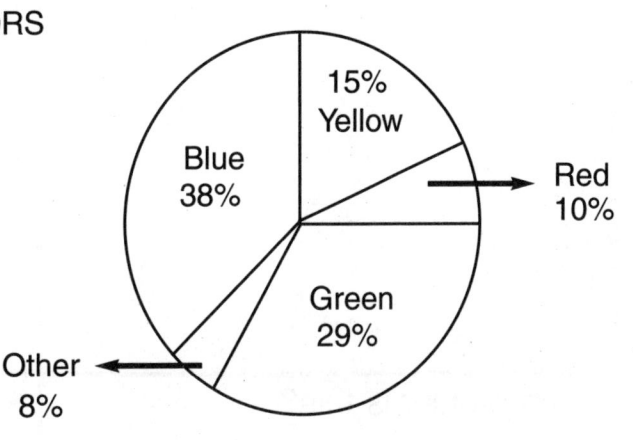

- Blue 38%
- Yellow 15%
- Red 10%
- Green 29%
- Other 8%

---

1. What color was the most popular with the people in this survey?

    a. red
    b. yellow
    c. blue
    d. green
    e. other

2. How many more people liked green than yellow?

    f. 8%
    g. 10%
    h. 15%
    j. 29%
    k. 14%

3. What percentage of people liked either yellow or red?

    a. 10%
    b. 15%
    c. 25%
    d. 29%
    e. 8%

4. What two groups make up exactly 67% of this graph?

    f. blue and yellow
    g. green and yellow
    h. yellow and red
    j. blue and green
    k. green and red

5. How many people voted for colors other than red, yellow, blue, and green?

    a. 8%
    b. 10%
    c. 15%
    d. 29%
    e. 38%

6. Which list is ordered from most to fewest votes?

    f. blue, green, other, yellow, red
    g. blue, green, yellow, red, other
    h. blue, green, red, other, yellow
    j. blue, green, red, yellow, other
    k. other, red, yellow, green, blue

© Carson-Dellosa CD-3754

Name _____

Skill: Application Practice

Directions
Read each question and choose the correct answer. Mark the space for the answer you have chosen. Mark NH if the answer is not here.

1. Beth bought 4 cans of cat food for $1.32. How much did each can cost?

   a. $0.33
   b. $0.40
   c. $0.32
   d. $1.00
   e. NH

5. Jason spent $25.00 on presents for 4 friends. How much did he spend on each person?

   a. $5.25
   b. $6.25
   c. $12.00
   d. $12.50
   e. NH

2. How much money is this?

   f. $0.72
   g. $1.21
   h. $0.61
   j. $1.13
   k. NH

6. Sarah earns $5.00 a day baby-sitting. How many days must she baby-sit to earn $40.00?

   f. 12 days
   g. 8 days
   h. 4 days
   j. 10 days
   k. NH

3. Look at the clock. It is p.m. What time will it be in 3 hours and 15 minutes?

   a. 5:45 p.m.
   b. 2:45 p.m.
   c. 6:15 p.m.
   d. 3:15 p.m.
   e. NH

7. Which shape is square?

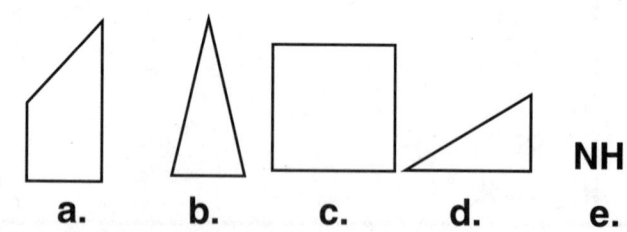

   a. b. c. d. NH
                        e.

4. David earned $10 on Monday, $12 on Tuesday and $14 on Wednesday. How much money did David earn in all?

   f. $22.00   j. $55.00
   g. $24.00   k. NH
   h. $36.00

8. What is the perimeter of this rectangle?

   3 ft
   36 ft

   f. 14 ft   j. 280 ft
   g. 24 ft   k. NH
   h. 28 ft

© Carson-Dellosa CD-3754

Name _____   Skill: Application Practice

Directions
Read each question and choose the correct answer. Mark the space for the answer you have chosen. Mark NH if the answer is not here.

---

1. Camille picked 8 tomatoes everyday for 6 days. How many tomatoes did she pick in all?

a. 14
b. 36
c. 48
d. 16
e. NH

2. Danielle puts 35 beads in each necklace she makes. How many beads does she need to make 4 necklaces?

f. 70
g. 140
h. 145
j. 120
k. NH

3. Carl has $25.00 in nickels. How many nickels does he have?

a. 5 nickels
b. 50 nickels
c. 500 nickels
d. 5,000 nickels
e. NH

4. What is the area of this square?

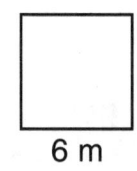

6 m

f. 14 sq m   j. 24 sq m
g. 36 sq m   k. NH
h. 49 sq m

5. Jamie collected 420 marbles. She gave half of them away. How many does she have left?

a. 240 marbles
b. 105 marbles
c. 120 marbles
d. 210 marbles
e. NH

6. Anna's purchase cost $ 7.29. How much change will she get from a ten-dollar bill?

f. $1.67
g. $2.71
h. $1.71
j. $2.77
k. NH

7. 1,000 grams =

a. 1,000 kilograms
b. 10 kilograms
c. 1 kilogram
d. 0.1
e. NH

8. What is the perimeter of this shape?

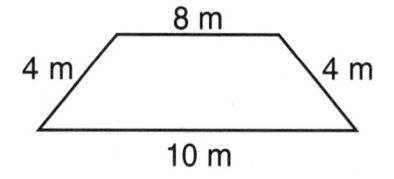

8 m
4 m   4 m
10 m

f. 14 ft   j. 280 ft
g. 24 ft   k. NH
h. 28 ft

© Carson-Dellosa CD-3754

Name _____   Skill: Application Practice

Directions
Read each question and choose the correct answer. Mark the space for the answer you have chosen. Mark NH if the answer is not here.

FAVORITE ANIMALS

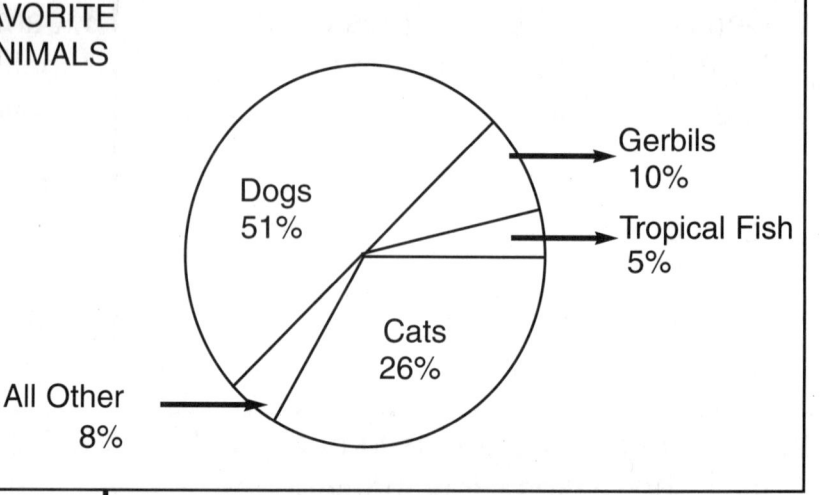

Beth asked 100 people to name their favorite animal. Here are the results of the survey.

- Dogs 51%
- Gerbils 10%
- Tropical Fish 5%
- Cats 26%
- All Other 8%

1. Which animal was the most popular with the people in this survey?

   a. cats
   b. dogs
   c. gerbils
   d. tropical fish
   e. other

2. What percentage of people preferred rabbits?

   f. 8%
   g. 10%
   h. 25%
   j. 29%
   k. NH

3. What percentage of people voted for either gerbils or tropical fish?

   a. 18%
   b. 15%
   c. 31%
   d. 29%
   e. NH

4. Which two groups make up exactly 31% of this graph?

   f. gerbils and other
   g. dogs and gerbils
   h. cats and tropical fish
   j. cats and gerbils
   k. NH

5. What percentage of people voted for animals other than dogs and cats?

   a. 8%
   b. 15%
   c. 23%
   d. 29%
   e. NH

6. Which list is ordered from most to fewest votes?

   f. other, cats, gerbils, fish, dogs
   g. dogs, cats, gerbils, fish, other
   h. cats, dogs, gerbils, fish, other
   j. dogs, cats, fish, gerbils, other
   k. NH

Name _____   Skill: Practice Test

Directions
Read each question and choose the correct answer. Mark the space for the answer you have chosen. Mark NH if the answer is not here.

---

1. What is the numeral for seven hundred thousand, three hundred thirty-nine?

   a. 700,339
   b. 703,039
   c. 7,339
   d. 70,339
   e. NH

5. Which group is in order from greatest to least?

   a. 41,023   40,123   41,302
   b. 40,123   41,023   41,302
   c. 41,023   41,302   40,123
   d. 41,302   41,023   40,123
   e. NH

---

2. What is the missing number?

   | 980 | 960 | 940 | | 900 |

   f. 939
   g. 920
   h. 910
   j. 901
   k. NH

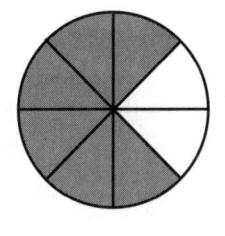

6. What fraction is shaded?

   f. $\frac{2}{4}$ j. $\frac{1}{3}$
   g. $\frac{5}{6}$ k. NH
   h. $\frac{2}{3}$

---

3. What number comes just <u>after</u> 3,899?

   a. 4,000
   b. 3,999
   c. 3,900
   d. 3,889
   e. NH

7. Which number sentence is true?

   a. 53 = 8 + 9
   b. 53 x 8 < 9
   c. 53 > 8 x 9
   d. 53 = 8 x 9
   e. NH

---

4. What is the value of the 8 in the number 153,869?

   f. hundreds
   g. thousands
   h. ten thousands
   j. hundred thousands
   k. NH

8. What is the missing number?

   12 x ☐ = 36

   f. 2
   g. 3
   h. 4
   j. 5
   k. NH

Name _____ Skill: Practice Test

Directions
Read each question and choose the correct answer. Mark the space for the answer you have chosen. Mark NH if the answer is not here.

1.

42
+ 51

a. 91
b. 13
c. 83
d. 93
e. NH

5.

610
− 137

a. 527
b. 473
c. 573
d. 483
e. NH

9.

7
× 6

a. 54
b. 48
c. 41
d. 39
e. NH

2.

76
− 14

f. 60
g. 62
h. 70
j. 52
k. NH

6.

2,243
+ 5,414

f. 7,757
g. 7,637
h. 7,657
j. 3,658
k. NH

10.

39
× 4

f. 156
g. 126
h. 394
j. 152
k. NH

3.

804
+ 578

a. 1,382
b. 1,282
c. 1,374
d. 1,222
e. NH

7.

8,295
+ 638

a. 8,823
b. 8,933
c. 8,833
d. 8,923
e. NH

11.

2)68̄

a. 22
b. 32
c. 34
d. 38
e. NH

4.

121
+ 458

f. 537
g. 577
h. 579
j. 589
k. NH

8.

7,096
− 4,258

f. 2,842
g. 3,838
h. 3,842
j. 2,838
k. NH

12.

7)490̄

f. 70
g. 60
h. 77
j. 67
k. NH

© Carson-Dellosa CD-3754

Name _____   Skill: Practice Test

Directions
Read each question and choose the correct answer. Mark the space for the answer you have chosen. Mark NH if the answer is not here.

The fourth graders at Johnston Middle School took a survey to see what pizza toppings were most popular. The graph shows the results of that survey.

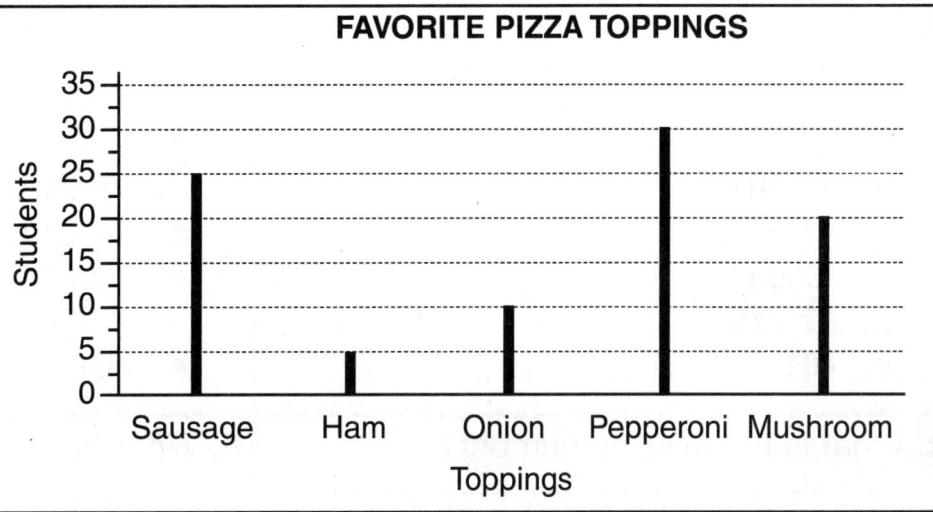

FAVORITE PIZZA TOPPINGS

(Bar graph: Sausage 25, Ham 5, Onion 10, Pepperoni 30, Mushroom 20)

Students / Toppings

---

1. What pizza topping was most popular with these students?

   a. pepperoni
   b. sausage
   c. mushroom
   d. onion
   e. NH

2. How many types of pizza toppings were selected in this survey?

   f. 4
   g. 5
   h. 6
   j. 7
   k. NH

3. What two toppings together got the same number of votes as pepperoni?

   a. onion and sausage
   b. ham and onion
   c. mushroom and onion
   d. sausage and mushroom
   e. NH

4. How many students liked ham or mushrooms?

   f. 40 students
   g. 20 students
   h. 15 students
   j. 25 students
   k. NH

5. How many students liked toppings that were a type of meat?

   a. 40
   b. 50
   c. 55
   d. 60
   e. NH

6. How many students took this survey?

   f. 85 students
   g. 90 students
   h. 95 students
   j. 100 students
   k. NH

© Carson-Dellosa CD-3754

Name _____ Skill: Practice Test

Directions
Read each question and choose the correct answer. Mark the space for the answer you have chosen. Mark NH if the answer is not here.

1. What is the numeral for seventeen thousand, five hundred one? a. 17,501 b. 17,500,001 c. 17,051 d. 17,510 e. NH	5. Which number has the greatest value? a. 78,349 b. 79,834 c. 89,743 d. 98,349 e. NH
2. What is the missing number? | 545 | 555 | 565 | 575 | f. 595 g. 585 h. 580 j. 576 k. NH	6. What fraction is less than $\frac{3}{9}$? f. $\frac{1}{2}$ g. $\frac{1}{3}$ h. $\frac{1}{4}$ j. $\frac{2}{3}$ k. NH
3. Which number is greater than 160,417? a. 106,580 b. 160,174 c. 141,659 d. 160,714 e. NH	7. What sign belongs in this statement? 121 ◯ 11 = 11 a. + b. − c. x d. ÷ e. NH
4. In what place value is the 6 in the number 269,253? f. hundreds g. thousands h. ten thousands j. hundred thousands k. NH	8. What means the same as 8,076? f. 8 + 0 + 7 + 6 g. 8,000 + 700 + 6 h. 8,000 + 70 + 6 j. 8,000 + 700 + 60 k. NH

© Carson-Dellosa CD-3754

Name _____   Skill: Practice Test

Directions
Read each question and choose the correct answer. Mark the space for the answer you have chosen. Mark NH if the answer is not here.

1. 2)140	a. 50 b. 60 c. 70 d. 80 e. NH	5. 709 − 155	a. 654 b. 504 c. 554 d. 604 e. NH	9. 7 × 8	a. 54 b. 56 c. 48 d. 63 e. NH
2. 571 − 328	f. 243 g. 257 h. 253 j. 143 k. NH	6. 3,639 + 5,188	f. 8,717 g. 8,727 h. 8,827 j. 8,817 k. NH	10. 76 × 4	f. 304 g. 284 h. 294 j. 314 k. NH
3. 670 + 258	a. 828 b. 928 c. 420 d. 820 e. NH	7. 8,354 + 621	a. 8,933 b. 8,375 c. 8,965 d. 8,365 e. NH	11. 6)240	a. 3 b. 4 c. 40 d. 30 e. NH
4. 4)176	f. 39 g. 41 h. 43 j. 44 k. NH	8. 12 × 8	f. 84 g. 96 h. 72 j. 108 k. NH	12. 9)738	f. 87 g. 81 h. 92 j. 82 k. NH

© Carson-Dellosa CD-3754

Name _____   Skill: Practice Test

Directions
Read each question and choose the correct answer. Mark the space for the answer you have chosen. Mark NH if the answer is not here.

Use the information given here to answer questions 1-3.

Jamie has $100 to spend at the Read More Book Store. She can choose from the following:
1) a hardback book for $9.35
2) a pack of comics for $11.60
3) a box of mysteries for $15.73
4) a stuffed animal with a book for $22.95

1. How much will it cost Jamie if she buys both the comics and mysteries?

   a. $20.95
   b. $25.08
   c. $27.33
   d. $32.30
   e. NH

2. What piece of information is <u>not</u> necessary to find out if Jamie can buy all the items on the list?

   f. the name of the book store
   g. the cost of the comics
   h. the cost of the mysteries
   j. how much money Jamie has
   k. NH

3. How much would it cost if Jamie bought all four choices?

   a. $58.53
   b. $59.63
   c. $60.45
   d. $62.39
   e. NH

4. Casey has to put 3 pencils in every bag. She has 56 bags. How many pencils does she need?

   f. 17 pencils
   g. 158 pencils
   h. 356 pencils
   j. 168 pencils
   k. NH

5. There were 2,354 people at the concert. 1,829 sat in seats and the rest stood. How many people stood?

   a. 4,183 people
   b. 1,525 people
   c. 525 people
   d. 635 people
   e. NH

6. Jerry can make 30 ornaments in one hour. At this rate, how long will it take him to make 900 ornaments?

   f. 27 hours
   g. 30 hours
   h. 3 hours
   j. 300 hours
   k. NH

7. Tom has 214 soccer cards, 392 baseball cards, and 117 hockey cards. How many cards does Tom have?

   a. 723 cards
   b. 331 cards
   c. 606 cards
   d. 509 cards
   e. NH

Name _____

Directions
Read each question and choose the correct answer. Mark the space for the answer you have chosen. Mark NH if the answer is not here.

Skill: Practice Test

1. What is the numeral for nine hundred eight thousand, nine hundred nine?

   a. 980,909
   b. 989,009
   c. 908,099
   d. 908,909
   e. NH

2. What is the missing number?

   | 21 | 28 | 35 | | 49 |

   f. 42
   g. 36
   h. 48
   j. 43
   k. NH

3. What is 24,851,099 rounded to the nearest hundred thousand?

   a. 25,000,000
   b. 24,950,000
   c. 24,900,000
   d. 24,851,000
   e. NH

4. What is the value of the 4 in the number 91,476,850?

   f. hundreds
   g. thousands
   h. ten thousands
   j. hundred thousands
   k. NH

5. Which number has the least value?

   a. 900,009
   b. 909,000
   c. 990,000
   d. 900,090
   e. NH

6. What is $55\frac{3}{100}$ written as a decimal?

   f. 0.553
   g. 55.03
   h. 5.53
   j. 55.3
   k. NH

7. What sign makes this statement true?

   75 − 25 ◯ 5 x 11

   a. >
   b. <
   c. =
   d. +
   e. NH

8. What is another way of saying this?

   5 x 8

   f. 8 + 8 + 8 + 8 + 8
   g. 8 ÷ 5
   h. 8 + 5
   j. 5 + 5 + 5 + 5 + 5
   k. NH

© Carson-Dellosa CD-3754   98

Name _____ Skill: Practice Test

Directions
Read each question and choose the correct answer. Mark the space for the answer you have chosen. Mark NH if the answer is not here.

1.
  34
 + 5

  a. 31
  b. 39
  c. 49
  d. 38
  e. NH

2.
  87
 − 16

  f. 71
  g. 73
  h. 91
  j. 72
  k. NH

3.
  264
 + 515

  a. 751
  b. 771
  c. 779
  d. 789
  e. NH

4.
  478
 + 53

  f. 531
  g. 521
  h. 425
  j. 431
  k. NH

5.
  711
 − 84

  a. 773
  b. 627
  c. 673
  d. 727
  e. NH

6.
  4,273
 + 3,104

  f. 7,177
  g. 7,307
  h. 7,377
  j. 7,371
  k. NH

7.
  6,647
 + 9,823

  a. 15,460
  b. 15,470
  c. 16,470
  d. 16,560
  e. NH

8.
  9,827
 − 5,956

  f. 3,131
  g. 3,871
  h. 3,831
  j. 3,861
  k. NH

9.
  12
 × 7

  a. 78
  b. 80
  c. 82
  d. 84
  e. NH

10.
  58
 × 9

  f. 452
  g. 522
  h. 598
  j. 958
  k. NH

11.
  8)‾200

  a. 25
  b. 50
  c. 15
  d. 22
  e. NH

12.
  6)‾312

  f. 62
  g. 60
  h. 52
  j. 50
  k. NH

© Carson-Dellosa CD-3754

Name _____

Skill: Practice Test

Directions
Read each question and choose the correct answer. Mark the space for the answer you have chosen. Mark NH if the answer is not here.

---

1. Which shape is symmetrical?

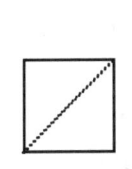

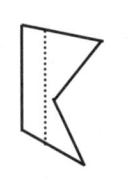

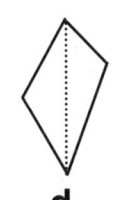

a.   b.   c.   d.   e. NH

2. Joe can type 12 pages in 30 minutes. How many pages can he type in 3 hours and 30 minutes?

f. 72 pages
g. 84 pages
h. 36 pages
j. 360 pages
k. NH

3. Which figure is a cone?

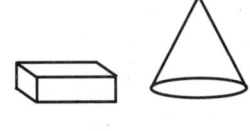

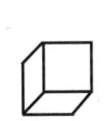

a.   b.   c.   d.   e. NH

4. What is the area of this rectangle?

f. 28 sq ft
g. 36 sq ft
h. 48 sq ft
j. 54 sq ft
k. NH

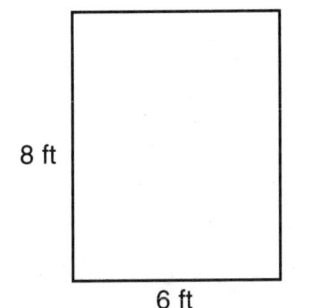

8 ft
6 ft

5. What is the perimeter of this rectangle?

a. 243 in
b. 72 in
c. 54 in
d. 36 in
e. NH

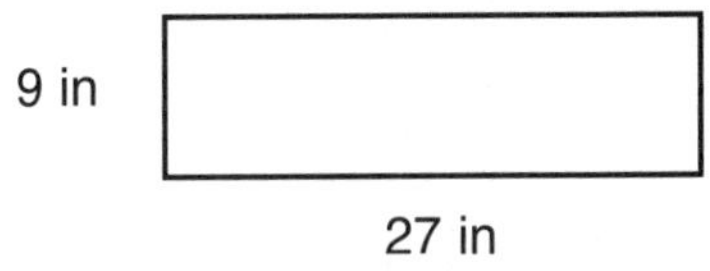

9 in
27 in

6. The minute hand is not shown on this clock. If the time is 8:40, at what number should the minute hand point?

f. 4
g. 8
h. 9
j. 10
k. NH

7. How much money is shown?

a. $2.12
b. $6.27
c. $6.31
d. $6.51
e. NH

8. 354 meters =

f. 3,540 kilometers
g. 35,400 millimeters
h. 35,400 centimeters
j. 3.54 kilometers
k. NH

© Carson-Dellosa CD-3754

Name _____   Skill: Practice Test

Directions
Read each question and choose the correct answer. Mark the space for the answer you have chosen. Mark NH if the answer is not here.

---

1. What is the numeral for forty three thousand, five hundred twelve?

   a. 435,012
   b. 430,512
   c. 43,512
   d. 4,312
   e. NH

5. Which group is in order from greatest to least?

   a. 31,023   30,123   31,302
   b. 30,123   31,023   31,302
   c. 31,023   31,302   30,123
   d. 31,302   31,023   30,123
   e. NH

---

2. What is the missing number?

| 612 | 610 | 608 | | 604 |

   f. 406
   g. 606
   h. 610
   j. 160
   k. NH

6. What number comes just <u>after</u> 2,989

   f. 2,999
   g. 2,988
   h. 2,990
   j. 3,000
   k. NH

---

3. What number comes just <u>after</u> 4,399?

   a. 4,000
   b. 4,400
   c. 3,304
   d. 3,499
   e. NH

7. Which number sentence is true?

   a. 64 = 8 + 9
   b. 64 x 8 < 9
   c. 64 > 8 x 9
   d. 64 = 8 x 8
   e. NH

---

4. What is the value of the 5 in the number 153,869?

   f. hundreds
   g. thousands
   h. ten thousands
   j. hundred thousands
   k. NH

8. What is the missing number?

12 x ☐ = 24

   f. 2
   g. 3
   h. 4
   j. 5
   k. NH

© Carson-Dellosa CD-3754   101

Name _____

Skill: Practice Test

Directions
Read each question and choose the correct answer. Mark the space for the answer you have chosen. Mark NH if the answer is not here.

1.
```
  32
+ 99
```
a. 132
b. 130
c. 121
d. 131
e. NH

5.
```
  610
- 227
```
a. 527
b. 473
c. 573
d. 383
e. NH

9.
```
   7
 x 8
```
a. 54
b. 48
c. 56
d. 39
e. NH

2.
```
  77
- 12
```
f. 29
g. 65
h. 72
j. 52
k. NH

6.
```
  1,999
+ 3,588
```
f. 5,587
g. 5,667
h. 7,668
j. 5,597
k. NH

10.
```
   39
 x  5
```
f. 156
g. 126
h. 394
j. 195
k. NH

3.
```
  810
+ 600
```
a. 1,010
b. 1,410
c. 1,140
d. 1,510
e. NH

7.
```
  8,295
+ 1,638
```
a. 9,823
b. 9,933
c. 9,833
d. 9,923
e. NH

11.
2)̄72
a. 22
b. 32
c. 36
d. 38
e. NH

4.
```
  121
+ 358
```
f. 437
g. 477
h. 479
j. 489
k. NH

8.
```
  5,096
- 4,258
```
f. 842
g. 838
h. 842
j. 1,838
k. NH

12.
7)̄560
f. 70
g. 90
h. 77
j. 67
k. NH

© Carson-Dellosa CD-3754

Name _____

Skill: Practice Test

Directions
Read each question and choose the correct answer. Mark the space for the answer you have chosen. Mark NH if the answer is not here.

The fifth graders at Fair Grove School took a survey to see which flavor ice cream was most popular. The graph shows the results of that survey.

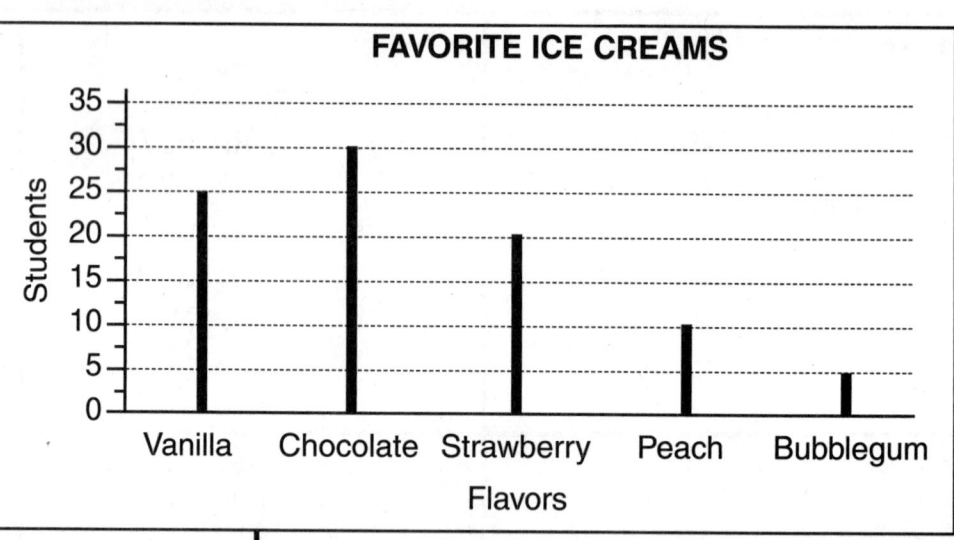

FAVORITE ICE CREAMS

(Bar graph showing Students vs Flavors: Vanilla = 25, Chocolate = 30, Strawberry = 20, Peach = 10, Bubblegum = 5)

1. Which ice cream was most popular with these students?

   a. vanilla
   b. chocolate
   c. strawberry
   d. bubblegum
   e. NH

2. How many flavors of ice cream were selected in this survey?

   f. 4
   g. 5
   h. 6
   j. 7
   k. NH

3. What two flavors together got the same number of votes as chocolate?

   a. vanilla and peach
   b. bubblegum and vanilla
   c. strawberry and peach
   d. peach and bubblegum
   e. NH

4. How many students liked bubblegum or vanilla?

   f. 40 students
   g. 20 students
   h. 15 students
   j. 25 students
   k. NH

5. How many students liked fruit flavors?

   a. 40
   b. 50
   c. 55
   d. 30
   e. NH

6. How many students took this survey?

   f. 85 students
   g. 90 students
   h. 95 students
   j. 100 students
   k. NH

© Carson-Dellosa CD-3754

Name _____

Skill: Practice Test

Directions
Read each question and choose the correct answer. Mark the space for the answer you have chosen. Mark NH if the answer is not here.

1. What is the numeral for twenty one thousand, three hundred four?

   a. 213,400
   b. 20,134
   c. 21,304
   d. 213,004
   e. NH

2. What is the missing number?

   | 345 | 355 | 365 | 375 | |

   f. 395
   g. 385
   h. 380
   j. 355
   k. NH

3. Which number is greater than 161,512?

   a. 106,580
   b. 161,521
   c. 141,659
   d. 160,714
   e. NH

4. In what place value is the 5 in the number 269,253?

   f. hundreds
   g. thousands
   h. ten thousands
   j. tens
   k. NH

5. Which number has the least value?

   a. 78,349
   b. 99,350
   c. 89,743
   d. 78,399
   e. NH

6. What fraction is more than $\frac{1}{2}$?

   f. $\frac{1}{5}$ g. $\frac{1}{3}$ h. $\frac{1}{4}$ j. $\frac{2}{3}$ k. NH

7. What sign belongs in this statement?

   144 ◯ 12 = 12

   a. +
   b. −
   c. x
   d. ÷
   e. NH

8. What means the same as 7,086?

   f. 7 + 0 + 7 + 6
   g. 7,000 + 800 + 6
   h. 7,000 + 80 + 6
   j. 7,000 + 700 + 60
   k. NH

104

Name _____

Directions

Read each question and choose the correct answer. Mark the space for the answer you have chosen. Mark NH if the answer is not here.

Skill: Practice Test

1.

2)120

a. 50
b. 60
c. 70
d. 80
e. NH

2.

600
− 328

f. 243
g. 257
h. 272
j. 143
k. NH

3.

670
+300

a. 370
b. 790
c. 970
d. 820
e. NH

4.

4)176

f. 39
g. 41
h. 43
j. 44
k. NH

5.

709
− 200

a. 654
b. 504
c. 509
d. 604
e. NH

6.

3,639
+ 6,001

f. 8,717
g. 8,740
h. 9,640
j. 8,817
k. NH

7.

8,354
+ 521

a. 8,833
b. 8,275
c. 8,865
d. 8,265
e. NH

8.

12
x 7

f. 84
g. 96
h. 72
j. 108
k. NH

9.

7
x 9

a. 54
b. 56
c. 48
d. 63
e. NH

10.

25
x 4

f. 304
g. 120
h. 100
j. 80
k. NH

11.

6)66

a. 33
b. 11
c. 10
d. NH
e.

12.

9)126

f. 90
g. 14
h. 16
j. 12
k. NH

© Carson-Dellosa CD-3754

Name _____

Skill: Practice Test

Directions
Read each question and choose the correct answer. Mark the space for the answer you have chosen. Mark NH if the answer is not here.

Use the information given here to answer questions 1-3.

Anna has $25 to spend at Parks Art Store. She can choose from the following:
1) colored pencils for $4.99
2) a watercolor and brush set for $5.99
3) a *How to Draw Birds* book for $9.99
4) an art supply kit for $14.95

1. How much will it cost Anna if she buys both the colored pencils and the watercolor set?

    a. $14.95
    b. $4.99
    c. $10.98
    d. $15.98
    e. NH

2. What piece of information is <u>not</u> necessary to find out if Anna can buy all the items on the list?

    f. the location of the store
    g. the cost of the colored pencils
    h. the cost of the art supply kit
    j. how much money Anna has
    k. NH

3. How much would it cost if Anna bought all four choices?

    a. $35.99
    b. $20.99
    c. $35.92
    d. $30.93
    e. NH

4. Susan's dog eats 2 cans of dog food each day. How many cans does the dog eat in two weeks?

    f. 14 cans
    g. 28 cans
    h. 7 cans
    j. 30 cans
    k. NH

5. 425 people each bought a $2.00 ticket to the basketball game. How much money was collected?

    a. $212.50
    b. $427.00
    c. $850.000
    d. $825.00
    e. NH

6. Steve did 15 minutes of homework in each of 4 subjects. How many minutes did he spend on homework?

    f. 90 minutes
    g. 60 minutes
    h. 60 hours
    j. 30 minutes
    k. NH

7. Joseph scored 14, 16, and 21 points in 3 different basketball games. How many points did he score in all?

    a. 81 points
    b. 30 points
    c. 51 points
    d. 82 points
    e. NH

Name _____

Skill: Practice Test

Directions
Read each question and choose the correct answer. Mark the space for the answer you have chosen. Mark NH if the answer is not here.

1. What is the numeral for two hundred four thousand, nine hundred two?

   a. 24,902
   b. 240,902
   c. 204,902
   d. 249,002
   e. NH

2. What is the missing number?

   | 20 | 26 | 32 | | 44 |

   f. 42
   g. 36
   h. 38
   j. 34
   k. NH

3. What is 24,851,099 rounded to the nearest hundred?

   a. 25,000,000
   b. 24,950,000
   c. 24,900,000
   d. 24,851,100
   e. NH

4. What is the value of the 7 in the number 91,476,850?

   f. hundreds
   g. thousands
   h. ten thousands
   j. hundred thousands
   k. NH

5. Which number has the least value?

   a. 411,114
   b. 414,111
   c. 441,111
   d. 411,141
   e. NH

6. What is $4\frac{3}{10}$ written as a decimal?

   f. 0.43
   g. 44.03
   h. 4.53
   j. 4.3
   k. NH

7. What sign makes this statement true?

   $$100 + 12 \bigcirc 56 + 56$$

   a. >
   b. <
   c. =
   d. +
   e. NH

8. What is another way of saying this?

   $$4 \times 7$$

   f. 4 + 4 + 4
   g. 4 ÷ 7
   h. 4 + 7
   j. 7 + 7 + 7 + 7
   k. NH

© Carson-Dellosa CD-3754

Name _____

Skill: Practice Test

Directions
Read each question and choose the correct answer. Mark the space for the answer you have chosen. Mark NH if the answer is not here.

1. 23 + 6	a. 31 b. 23 c. 29 d. 38 e. NH	5. 750 − 83	a. 833 b. 627 c. 667 d. 766 e. NH	9. 12 × 10	a. 122 b. 120 c. 22 d. 220 e. NH
2. 90 − 15	f. 71 g. 105 h. 75 j. 26 k. NH	6. 4,107 3,199	f. 7,177 g. 7,307 h. 7,306 j. 7,371 k. NH	10. 60 × 9	f. 51 g. 150 h. 540 j. 450 k. NH
3. 231 + 112	a. 403 b. 433 c. 343 d. 123 e. NH	7. 5,299 + 1,004	a. 4,205 b. 4,295 c. 6,303 d. 5,303 e. NH	11. 8)384	a. 25 b. 48 c. 15 d. 84 e. NH
4. 522 + 53	f. 603 g. 469 h. 755 j. 575 k. NH	8. 655 − 589	f. 1,244 g. 67 h. 66 j. 99 k. NH	12. 6)534	f. 66 g. 23 h. 99 j. 89 k. NH

© Carson-Dellosa CD-3754

Name _____

Skill: Practice Test

Directions
Read each question and choose the correct answer. Mark the space for the answer you have chosen. Mark NH if the answer is not here.

1. Which shape is square?

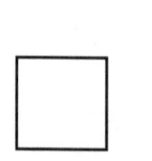

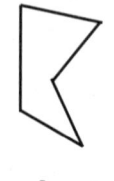

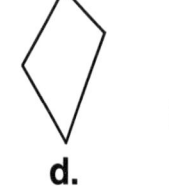

a.   b.   c.   d.   NH
              e.

2. Sheila can wrap one package every 10 minutes. How long will it take her to wrap 15 packages?

f. 150 minutes
g. 25 minutes
h. 100 minutes
j. 360 minutes
k. NH

3. Which figure is a cylinder?

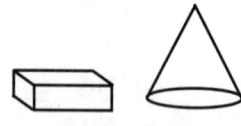

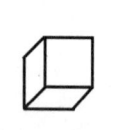

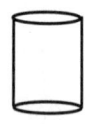

a.   b.   c.   d.   NH
              e.

4. What is the area of this rectangle?

f. 28 sq ft
g. 25 sq ft
h. 15 sq ft
j. 54 sq ft
k. NH

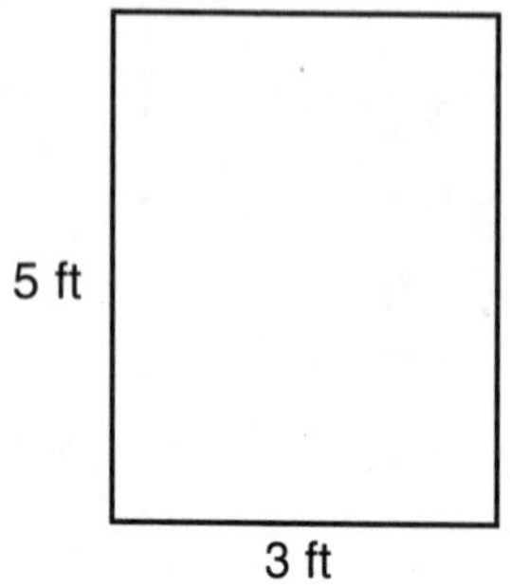

5 ft
3 ft

5. What is the perimeter of this rectangle?

a. 18 in
b. 72 in
c. 64 in
d. 32 in
e. NH

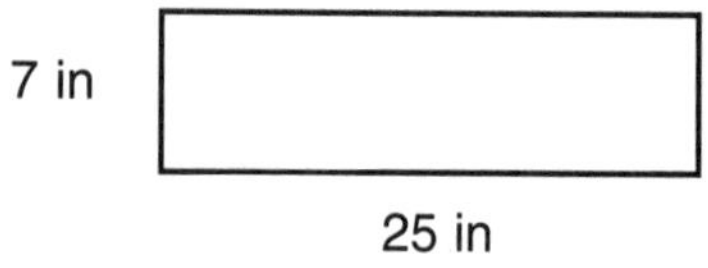

7 in
25 in

6. The hands are not shown on this clock. If the time is 2:40, at what number should the minute hand point?

f. 4
g. 8
h. 9
j. 2
k. NH

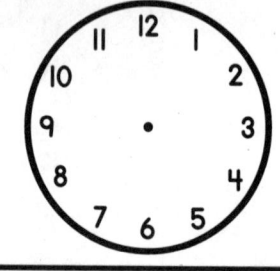

7. How much money is shown?

a. $3.11
b. $3.26
c. $6.03
d. $3.06
e. NH

8. 1 kilometer =

f. 10,000 meters
g. 1,000 meters
h. 100 centimeters
j. 10 meters
k. NH

© Carson-Dellosa CD-3754

Name _____   Skill: Practice Test

Directions
Read each question and choose the correct answer. Mark the space for the answer you have chosen. Mark NH if the answer is not here.

1. 33 + 50	a. 91 b. 13 c. 83 d. 93 e. NH	5. 510 − 137	a. 527 b. 473 c. 573 d. 373 e. NH	9. 8 × 6	a. 54 b. 48 c. 41 d. 39 e. NH
2. 76 − 10	f. 60 g. 62 h. 66 j. 52 k. NH	6. 2,243 + 2,202	f. 7,757 g. 4,544 h. 4,445 j. 41 k. NH	10. 40 × 4	f. 156 g. 126 h. 160 j. 80 k. NH
3. 950 + 120	a. 830 b. 1,300 c. 1,700 d. 1,070 e. NH	7. 120 + 220	a. 322 b. 220 c. 340 d. 430 e. NH	11. 2)‾72	a. 22 b. 32 c. 34 d. 38 e. NH
4. 121 + 320	f. 202 g. 440 h. 441 j. 344 k. NH	8. 100 − 42	f. 28 g. 68 h. 58 j. 85 k. NH	12. 7)‾623	f. 70 g. 89 h. 88 j. 87 k. NH

© Carson-Dellosa CD-3754   110

Name _____

Directions
Read each question and choose the correct answer. Mark the space for the answer you have chosen. Mark NH if the answer is not here.

Skill: Practice Test

---

1. What is the numeral for fourteen thousand, one?

   a. 14,001
   b. 14,100
   c. 14,010
   d. 1,401
   e. NH

---

5. Which group is in order from greatest to least?

   a. 81,043   80,143   81,304
   b. 80,143   81,043   81,304
   c. 81,043   81,304   80,143
   d. 81,304   81,043   80,143
   e. NH

---

2. What is the missing number?

   | 640 | 650 | 660 | | 680 |

   f. 675
   g. 670
   h. 665
   j. 600
   k. NH

---

6. What fraction is <u>not</u> shaded?

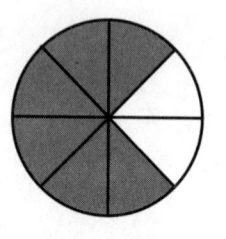

   f. $\frac{2}{4}$ j. $\frac{1}{3}$
   g. $\frac{2}{8}$ k. NH
   h. $\frac{2}{3}$

---

3. What number comes just <u>after</u> 1,009?

   a. 1,100
   b. 1,010
   c. 1,008
   d. 2,009
   e. NH

---

7. Which number sentence is true?

   a. 48 = 6 + 8
   b. 48 x 6 < 8
   c. 48 > 6 x 8
   d. 48 = 6 x 8
   e. NH

---

4. What is the value of the 1 in the number 153,869?

   f. hundreds
   g. thousands
   h. ten thousands
   j. hundred thousands
   k. NH

---

8. What is the missing number?

   14 x ☐ = 56

   f. 2
   g. 3
   h. 4
   j. 5
   k. NH

© Carson-Dellosa CD-3754   111

Name _____

Skill: Practice Test

Directions
Read each question and choose the correct answer. Mark the space for the answer you have chosen. Mark NH if the answer is not here.

---

1. What is the numeral for sixteen thousand, five hundred two?

a. 16,502
b. 16,500,002
c. 16,052
d. 16,210
e. NH

2. What is the missing number?

| 410 | 420 | 430 | 440 | |

f. 540
g. 450
h. 430
j. 45
k. NH

3. Which number is greater than 150,517?

a. 106,580
b. 150,174
c. 141,659
d. 150,714
e. NH

4. In what place value is the 9 in the number 269,253?

f. hundreds
g. thousands
h. ten thousands
j. hundred thousands
k. NH

5. Which number has the greatest value?

a. 61,349
b. 79,221
c. 45,743
d. 79,349
e. NH

6. What fraction is less than $\frac{1}{9}$?

f. $\frac{1}{2}$ g. $\frac{1}{3}$ h. $\frac{1}{4}$ j. $\frac{2}{3}$ k. NH

7. What sign belongs in this statement?

100 ◯ 10 = 10

a. +
b. −
c. ×
d. ÷
e. NH

8. What means the same as 7,076?

f. 7 + 0 + 8 + 6
g. 7,000 + 800 + 6
h. 7,000 + 70 + 6
j. 7,000 + 800 + 60
k. NH

Name _____   Skill: Practice Test

Directions
Read each question and choose the correct answer. Mark the space for the answer you have chosen. Mark NH if the answer is not here.

1. 2)120	a. 50 b. 60 c. 70 d. 80 e. NH	5. 850 − 255	a. 654 b. 504 c. 595 d. 604 e. NH	9. 7 x 4	a. 54 b. 28 c. 48 d. 11 e. NH
2. 89 − 21	f. 67 g. 110 h. 68 j. 86 k. NH	6. 78 + 21	f. 47 g. 57 h. 100 j. 99 k. NH	10. 75 x 4	f. 309 g. 79 h. 300 j. 314 k. NH
3. 650 + 358	a. 292 b. 1,250 c. 1,002 d. 1,008 e. NH	7. 250 + 621	a. 781 b. 871 c. 420 d. 891 e. NH	11. 5)250	a. 3 b. 5 c. 50 d. 30 e. NH
4. 4)160	f. 39 g. 41 h. 40 j. 44 k. NH	8. 16 x 8	f. 24 g. 104 h. 72 j. 128 k. NH	12. 8)440	f. 22 g. 55 h. 45 j. 65 k. NH

© Carson-Dellosa CD-3754   113

Name _____   Skill: Practice Test

Directions
Read each question and choose the correct answer. Mark the space for the answer you have chosen. Mark NH if the answer is not here.

1. What is the numeral for forty-two thousand, three hundred fourteen?

   a. 42,414
   b. 42,314
   c. 423,014
   d. 42,340
   e. NH

2. What is the missing number?

   | 21 | 28 | 35 | | 49 |

   f. 42
   g. 36
   h. 48
   j. 43
   k. NH

3. What is 24,851,099 rounded to the nearest hundred thousand?

   a. 25,000,000
   b. 24,950,000
   c. 24,900,000
   d. 24,851,000
   e. NH

4. What is the value of the 4 in the number 91,476,850?

   f. hundreds
   g. thousands
   h. ten thousands
   j. hundred thousands
   k. NH

5. Which number has the least value?

   a. 700,007
   b. 709,000
   c. 790,000
   d. 700,070
   e. NH

6. What is $2\frac{9}{10}$ written as a decimal?

   f. 0.229
   g. 2.019
   h. 2.910
   j. 2.9
   k. NH

7. What sign makes this statement true?

   125 − 25 ◯ 25 + 100

   a. >
   b. <
   c. =
   d. +
   e. NH

8. What is another way of saying this?

   3 x 6

   f. 3 + 6 + 6
   g. 6 ÷ 3
   h. 6 + 3
   j. 6 + 6 + 6
   k. NH

© Carson-Dellosa CD-3754

Name _____

Skill: Practice Test

Directions
Read each question and choose the correct answer. Mark the space for the answer you have chosen. Mark NH if the answer is not here.

---

1. Which shape is circular?

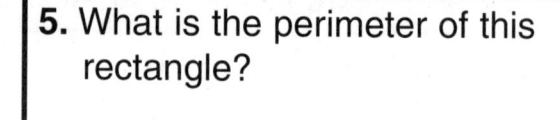

a.   b.   c.   d.   NH
                    e.

2. Tom can skate around the rink 3 times each minute. How many times can he skate around the rink in 16 minutes?

f. 19 times
g. 48 times
h. 13 times
j. 12 times
k. NH

3. Which figure is rectangular?

a.   b.   c.   d.   NH
                    e.

4. What is the area of this rectangle?

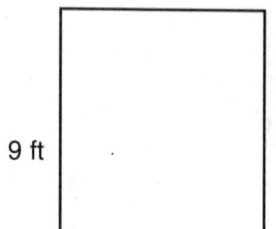

f. 28 sq ft
g. 45 sq ft
h. 48 sq ft
j. 14 sq ft
k. NH

9 ft
5 ft

---

5. What is the perimeter of this rectangle?

a. 114 in
b. 70 in
c. 19 in
d. 38 in
e. NH

5 in
14 in

6. The hands are not shown on this clock. If the time is 11:30, at what number should the minute hand point?

f. 11
g. 6
h. 9
j. 12
k. NH

7. How much money is shown?

a. $2.73
b. $2.63
c. $3.50
d. $3.03
e. NH

8. 10 millimeters =

f. 0.1 centimeters
g. 1 centimeter
h. 10 centimeters
j. 100 centimeters
k. NH

© Carson-Dellosa CD-3754

115

Answer Key

Page 1	Page 2	Page 3	Page 4	Page 5	Page 6
1. d	1. a	1. c	1. a	1. b	1. d
2. h	2. h	2. h	2. f	2. h	2. g
3. d	3. c	3. b	3. a	3. c	3. d
4. g	4. g	4. f	4. k	4. g	4. f
5. b	5. d	5. a	5. c	5. b	5. c
6. g	6. g	6. k	6. g	6. f	6. h
7. d	7. c	7. b	7. d	7. d	7. b
8. h	8. f	8. f	8. h	8. h	8. k

Page 7	Page 8	Page 9	Page 10	Page 11	Page 12
1. c	1. b	1. d	1. d	1. d	1. c
2. f	2. j	2. f	2. g	2. h	2. f
3. b	3. c	3. c	3. b	3. a	3. d
4. g	4. j	4. h	4. k	4. g	4. f
5. d	5. c	5. c	5. a	5. k	5. e
6. g	6. k	6. h	6. h	6. k	6. f
7. a	7. b	7. b	7. d	7. d	7. e
8. h	8. g	8. j	8. h	8. j	8. g

Page 13	Page 14	Page 15	Page 16	Page 17	Page 18
1. c	1. c	1. c	1. a	1. c	1. a
2. g	2. f	2. j	2. g	2. f	2. g
3. d	3. b	3. c	3. c	3. b	3. a
4. h	4. h	4. f	4. j	4. j	4. h
5. d	5. e	5. d	5. b	5. d	5. d
6. h	6. j	6. f	6. k	6. k	6. j
7. d	7. b	7. c	7. a	7. b	7. b
8. h	8. g	8. k	8. g	8. k	8. g

Page 19	Page 20	Page 21	Page 22	Page 23	Page 24
1. e	1. d	1. b	1. c	1. b	1. c
2. f	2. g	2. k	2. k	2. g	2. k
3. c	3. b	3. b	3. b	3. d	3. a
4. j	4. j	4. j	4. k	4. f	4. k
5. c	5. b	5. d	5. a	5. d	5. d
6. g	6. k	6. k	6. f	6. f	6. f
7. a	7. c	7. a	7. d	7. c	7. a
8. j	8. j	8. h	8. h	8. k	8. h

Answer Key

Page 25
1. c
2. f
3. a
4. j
5. b
6. h
7. d
8. h

Page 26
1. b
2. k
3. c
4. j
5. d
6. k
7. b
8. g

Page 27
1. a
2. g
3. d
4. g
5. c
6. j
7. b
8. g

Page 28
1. c
2. g
3. d
4. f
5. c
6. f
7. c
8. g

Page 29
1. a
2. g
3. d
4. f
5. b
6. f
7. d
8. h

Page 30
1. c
2. h
3. b
4. f
5. e
6. g
7. d
8. h

Page 31
1. c
2. h
3. b
4. j
5. c
6. k
7. a
8. g
9. b
10. j
11. e
12. g

Page 32
1. d
2. j
3. e
4. h
5. b
6. f
7. b
8. h
9. e
10. f
11. b
12. f

Page 33
1. c
2. f
3. b
4. f
5. a
6. g
7. c
8. h
9. b
10. j
11. c
12. g

Page 34
1. d
2. f
3. a
4. k
5. b
6. h
7. c
8. j
9. a
10. h
11. b
12. g

Page 35
1. e
2. g
3. e
4. h
5. a
6. j
7. c
8. g
9. d
10. j
11. b
12. f

Page 36
1. b
2. j
3. c
4. h
5. b
6. h
7. d
8. h
9. a
10. g
11. b
12. f

Page 37
1. c
2. f
3. a
4. j
5. b
6. h
7. e
8. h
9. e
10. j
11. c
12. g

Page 38
1. b
2. h
3. a
4. h
5. b
6. h
7. d
8. g
9. b
10. f
11. b
12. j

Page 39
1. c
2. j
3. b
4. k
5. c
6. j
7. c
8. k
9. b
10. f
11. e
12. g

Page 40
1. a
2. g
3. a
4. h
5. b
6. g
7. a
8. j
9. a
10. h
11. d
12. g

Page 41
1. d
2. g
3. a
4. f
5. c
6. f
7. d
8. j
9. a
10. h
11. b
12. h

Page 42
1. a
2. g
3. a
4. g
5. d
6. j
7. e
8. h
9. d
10. f
11. c
12. h

Answer Key

Page 43	Page 44	Page 45	Page 46	Page 47	Page 48
1. b	1. c	1. c	1. b	1. a	1. a
2. g	2. j	2. h	2. f	2. j	2. j
3. d	3. f	3. b	3. c	3. c	3. d
4. h	4. f	4. g	4. g	4. g	4. f
5. a	5. a	5. e	5. c	5. d	5. e
6. g	6. f	6. g	6. g	6. f	6. k
7. a	7. c	7. e	7. d	7. a	7. a
8. g	8. h	8. k	8. j	8. h	8. f
9. a	9. d	9. b	9. e	9. c	9. c
10. h	10. f	10. k	10. f	10. k	10. k
11. b	11. d	11. c	11. c	11. d	11. d
12. j	12. g	12. j	12. j	12. k	12. g

Page 49	Page 50	Page 51	Page 52	Page 53	Page 54
1. d	1. c	1. b	1. d	1. b	1. a
2. f	2. g	2. k	2. j	2. j	2. g
3. a	3. a	3. b	3. a	3. a	3. e
4. g	4. h	4. f	4. h	4. g	4. g
5. c	5. c	5. d	5. b	5. c	5. a
6. g	6. j	6. f	6. j	6. g	6. k
7. e	7. b	7. c	7. b	7. d	7. a
8. h	8. g	8. f	8. j	8. f	8. k
9. a	9. a	9. d	9. e	9. c	9. b
10. g	10. j	10. j	10. g	10. j	10. g
11. a	11. c	11. b	11. b	11. b	11. b
12. g	12. f	12. g	12. h	12. g	12. j

Page 55	Page 56	Page 57	Page 58	Page 59	Page 60
1. b	1. b	1. b	1. c	1. a	1. c
2. g	2. j	2. h	2. f	2. h	2. g
3. a	3. a	3. c	3. e	3. c	3. c
4. f	4. f	4. j	4. g	4. h	4. g
5. d	5. b	5. c	5. d	5. e	5. d
6. h	6. h	6. j	6. k	6. g	6. k
7. c	7. c	7. c	7. a	7. d	7. a
8. k	8. j	8. g	8. j	8. g	8. k

Answer Key

Page 61	Page 62	Page 63	Page 64	Page 65	Page 66
1. b	1. b	1. d	1. b	1. c	1. a
2. f	2. j	2. h	2. g	2. g	2. f
3. c	3. c	3. b	3. b	3. b	3. d
4. h	4. f	4. k	4. j	4. h	4. g
5. d	5. c	5. d	5. c	5. a	5. c
6. h	6. k	6. f	6. k	6. j	6. g
7. b	7. b	7. c	7. b	7. d	7. d
8. g	8. k	8. j	8. g	8. g	8. k
9. a	9. a		9. b	9. d	
10. g	10. k		10. g	10. g	
11. d	11. b		11. d	11. a	
12. h	12. h		12. h	12. g	

Page 67	Page 68	Page 69	Page 70	Page 71	Page 72
1. d	1. b	1. d	1. a	1. c	1. a
2. j	2. f	2. h	2. g	2. h	2. g
3. e	3. c	3. c	3. d	3. g	3. a
4. h	4. k	4. h	4. h	4. f	4. g
5. b	5. c	5. a	5. c	5. g	5. c
6. f	6. k	6. j	6. f	6. j	6. f
7. a	7. c	7. b	7. a		7. b
8. g	8. g	8. j	8. j		8. j

Page 73	Page 74	Page 75	Page 76	Page 77	Page 78
1. d	1. d	1. c	1. b	1. c	1. b
2. g	2. k	2. g	2. f	2. k	2. f
3. e	3. b	3. e	3. b	3. b	3. c
4. f	4. k	4. j	4. j	4. j	4. j
5. b	5. c	5. c	5. c	5. c	5. f
6. k	6. k	6. j	6. j	6. k	6. j
7. b	7. c	7. d	7. c	7. f	7. b
8. j	8. g	8. g	8. h	8. h	8. j

Answer Key

Page 79	Page 80	Page 81	Page 82	Page 83	Page 84
1. d	1. c	1. b	1. c	1. a	1. b
2. h	2. g	2. j	2. f	2. h	2. f
3. d	3. d	3. a	3. d	3. d	3. e
4. g	4. h	4. h	4. g	4. h	4. h
5. b	5. c	5. b	5. c	5. c	5. d
6. h	6. g	6. g	6. f	6. j	6. g
7. d	7. f	7. c		7. e	
8. g	8. g	8. f		8. j	

Page 85	Page 86	Page 87	Page 88	Page 89	Page 90
1. d	1. b	1. d	1. c	1. f	1. c
2. h	2. j	2. g	2. k	2. f	2. g
3. b	3. f	3. c	3. c	3. a	3. a
4. g	4. j	4. j	4. j	4. h	4. g
5. c	5. b	5. c	5. f	5. b	5. d
6. j	6. g	6. f	6. g	6. g	6. f
	7. b	7. f		7. c	7. d
	8. h	8. k		8. k	8. k

Page 91	Page 92	Page 93	Page 94	Page 96	Page 97
1. b	1. f	1. d	1. f	1. c	1. c
2. k	2. g	2. h	2. g	2. f	2. f
3. b	3. c	3. b	3. c	3. b	3. e
4. h	4. g	4. f	4. k	4. h	4. j
5. d	5. d	5. b	5. d	5. c	5. c
6. k	6. k	6. j	6. j	6. g	6. f
	7. e	7. c	Page 95	7. e	7. b
	8. g	8. k	1. f	8. g	
		9. e	2. g	9. b	
		10. f	3. d	10. f	
		11. c	4. h	11. c	
		12. f	5. d	12. j	
			6. j		
			7. d		
			8. j		

Answer Key

Page 98
1. d
2. f
3. c
4. k
5. a
6. g
7. b
8. f

Page 99
1. b
2. f
3. c
4. f
5. b
6. h
7. c
8. g
9. d
10. g
11. a
12. h

Page 100
1. a
2. g
3. b
4. k
5. b
6. g
7. d
8. h

Page 101
1. a
2. g
3. a
4. h
5. d
6. h
7. d
8. f

Page 102
1. d
2. f
3. c
4. h
5. a
6. f
7. b
8. g

Page 103
1. b
2. f
3. c
4. k
5. d
6. g
7. b
8. g
9. c
10. j
11. c
12. k

Page 104
1. c
2. g
3. b
4. j
5. a
6. j
7. d
8. h

Page 105
1. b
2. h
3. a
4. j
5. c
6. h
7. k
8. f
9. e
10. h
11. b
12. g

Page 106
1. c
2. f
3. c
4. g
5. c
6. f
7. d

Page 107
1. c
2. h
3. e
4. g
5. f
6. j
7. c
8. k

Page 108
1. c
2. h
3. d
4. j
5. b
6. j
7. b
8. f
9. b
10. h
11. b
12. k

Page 109
1. a
2. f
3. e
4. h
5. c
6. j
7. d
8. g

Page 110
1. c
2. h
3. d
4. j
5. e
6. j
7. c
8. h
9. b
10. f
11. e
12. g

Page 111
1. a
2. f
3. b
4. j
5. e
6. g
7. g
8. h

Page 112
1. a
2. g
3. b
4. g
5. d
6. k
7. e
8. h

Page 113
1. b
2. g
3. d
4. h
5. b
6. h
7. b
8. f
9. b
10. f
11. c
12. g

Page 114
1. b
2. f
3. c
4. j
5. b
6. j
7. b
8. j

Page 115
1. d
2. f
3. a
4. g
5. d
6. g
7. b
8. f

Answer Sheet

Name _____

Page _____ Score _____	Page _____ Score _____	Page _____ Score _____
1. ⓐ ⓑ ⓒ ⓓ ⓔ 2. ⓕ ⓖ ⓗ ⓙ ⓚ 3. ⓐ ⓑ ⓒ ⓓ ⓔ 4. ⓕ ⓖ ⓗ ⓙ ⓚ 5. ⓐ ⓑ ⓒ ⓓ ⓔ 6. ⓕ ⓖ ⓗ ⓙ ⓚ 7. ⓐ ⓑ ⓒ ⓓ ⓔ 8. ⓕ ⓖ ⓗ ⓙ ⓚ 9. ⓐ ⓑ ⓒ ⓓ ⓔ 10. ⓕ ⓖ ⓗ ⓙ ⓚ 11. ⓐ ⓑ ⓒ ⓓ ⓔ 12. ⓕ ⓖ ⓗ ⓙ ⓚ	1. ⓐ ⓑ ⓒ ⓓ ⓔ 2. ⓕ ⓖ ⓗ ⓙ ⓚ 3. ⓐ ⓑ ⓒ ⓓ ⓔ 4. ⓕ ⓖ ⓗ ⓙ ⓚ 5. ⓐ ⓑ ⓒ ⓓ ⓔ 6. ⓕ ⓖ ⓗ ⓙ ⓚ 7. ⓐ ⓑ ⓒ ⓓ ⓔ 8. ⓕ ⓖ ⓗ ⓙ ⓚ 9. ⓐ ⓑ ⓒ ⓓ ⓔ 10. ⓕ ⓖ ⓗ ⓙ ⓚ 11. ⓐ ⓑ ⓒ ⓓ ⓔ 12. ⓕ ⓖ ⓗ ⓙ ⓚ	1. ⓐ ⓑ ⓒ ⓓ ⓔ 2. ⓕ ⓖ ⓗ ⓙ ⓚ 3. ⓐ ⓑ ⓒ ⓓ ⓔ 4. ⓕ ⓖ ⓗ ⓙ ⓚ 5. ⓐ ⓑ ⓒ ⓓ ⓔ 6. ⓕ ⓖ ⓗ ⓙ ⓚ 7. ⓐ ⓑ ⓒ ⓓ ⓔ 8. ⓕ ⓖ ⓗ ⓙ ⓚ 9. ⓐ ⓑ ⓒ ⓓ ⓔ 10. ⓕ ⓖ ⓗ ⓙ ⓚ 11. ⓐ ⓑ ⓒ ⓓ ⓔ 12. ⓕ ⓖ ⓗ ⓙ ⓚ
Page _____ Score _____	Page _____ Score _____	Page _____ Score _____
1. ⓐ ⓑ ⓒ ⓓ ⓔ 2. ⓕ ⓖ ⓗ ⓙ ⓚ 3. ⓐ ⓑ ⓒ ⓓ ⓔ 4. ⓕ ⓖ ⓗ ⓙ ⓚ 5. ⓐ ⓑ ⓒ ⓓ ⓔ 6. ⓕ ⓖ ⓗ ⓙ ⓚ 7. ⓐ ⓑ ⓒ ⓓ ⓔ 8. ⓕ ⓖ ⓗ ⓙ ⓚ 9. ⓐ ⓑ ⓒ ⓓ ⓔ 10. ⓕ ⓖ ⓗ ⓙ ⓚ 11. ⓐ ⓑ ⓒ ⓓ ⓔ 12. ⓕ ⓖ ⓗ ⓙ ⓚ	1. ⓐ ⓑ ⓒ ⓓ ⓔ 2. ⓕ ⓖ ⓗ ⓙ ⓚ 3. ⓐ ⓑ ⓒ ⓓ ⓔ 4. ⓕ ⓖ ⓗ ⓙ ⓚ 5. ⓐ ⓑ ⓒ ⓓ ⓔ 6. ⓕ ⓖ ⓗ ⓙ ⓚ 7. ⓐ ⓑ ⓒ ⓓ ⓔ 8. ⓕ ⓖ ⓗ ⓙ ⓚ 9. ⓐ ⓑ ⓒ ⓓ ⓔ 10. ⓕ ⓖ ⓗ ⓙ ⓚ 11. ⓐ ⓑ ⓒ ⓓ ⓔ 12. ⓕ ⓖ ⓗ ⓙ ⓚ	1. ⓐ ⓑ ⓒ ⓓ ⓔ 2. ⓕ ⓖ ⓗ ⓙ ⓚ 3. ⓐ ⓑ ⓒ ⓓ ⓔ 4. ⓕ ⓖ ⓗ ⓙ ⓚ 5. ⓐ ⓑ ⓒ ⓓ ⓔ 6. ⓕ ⓖ ⓗ ⓙ ⓚ 7. ⓐ ⓑ ⓒ ⓓ ⓔ 8. ⓕ ⓖ ⓗ ⓙ ⓚ 9. ⓐ ⓑ ⓒ ⓓ ⓔ 10. ⓕ ⓖ ⓗ ⓙ ⓚ 11. ⓐ ⓑ ⓒ ⓓ ⓔ 12. ⓕ ⓖ ⓗ ⓙ ⓚ

© Carson-Dellosa CD-3754

Answer Sheet

Name _____

Page _____	Page _____	Page _____
Score _____	Score _____	Score _____
1. ⓐ ⓑ ⓒ ⓓ ⓔ	1. ⓐ ⓑ ⓒ ⓓ ⓔ	1. ⓐ ⓑ ⓒ ⓓ ⓔ
2. ⓕ ⓖ ⓗ ⓙ ⓚ	2. ⓕ ⓖ ⓗ ⓙ ⓚ	2. ⓕ ⓖ ⓗ ⓙ ⓚ
3. ⓐ ⓑ ⓒ ⓓ ⓔ	3. ⓐ ⓑ ⓒ ⓓ ⓔ	3. ⓐ ⓑ ⓒ ⓓ ⓔ
4. ⓕ ⓖ ⓗ ⓙ ⓚ	4. ⓕ ⓖ ⓗ ⓙ ⓚ	4. ⓕ ⓖ ⓗ ⓙ ⓚ
5. ⓐ ⓑ ⓒ ⓓ ⓔ	5. ⓐ ⓑ ⓒ ⓓ ⓔ	5. ⓐ ⓑ ⓒ ⓓ ⓔ
6. ⓕ ⓖ ⓗ ⓙ ⓚ	6. ⓕ ⓖ ⓗ ⓙ ⓚ	6. ⓕ ⓖ ⓗ ⓙ ⓚ
7. ⓐ ⓑ ⓒ ⓓ ⓔ	7. ⓐ ⓑ ⓒ ⓓ ⓔ	7. ⓐ ⓑ ⓒ ⓓ ⓔ
8. ⓕ ⓖ ⓗ ⓙ ⓚ	8. ⓕ ⓖ ⓗ ⓙ ⓚ	8. ⓕ ⓖ ⓗ ⓙ ⓚ
9. ⓐ ⓑ ⓒ ⓓ ⓔ	9. ⓐ ⓑ ⓒ ⓓ ⓔ	9. ⓐ ⓑ ⓒ ⓓ ⓔ
10. ⓕ ⓖ ⓗ ⓙ ⓚ	10. ⓕ ⓖ ⓗ ⓙ ⓚ	10. ⓕ ⓖ ⓗ ⓙ ⓚ
11. ⓐ ⓑ ⓒ ⓓ ⓔ	11. ⓐ ⓑ ⓒ ⓓ ⓔ	11. ⓐ ⓑ ⓒ ⓓ ⓔ
12. ⓕ ⓖ ⓗ ⓙ ⓚ	12. ⓕ ⓖ ⓗ ⓙ ⓚ	12. ⓕ ⓖ ⓗ ⓙ ⓚ

Page _____	Page _____	Page _____
Score _____	Score _____	Score _____
1. ⓐ ⓑ ⓒ ⓓ ⓔ	1. ⓐ ⓑ ⓒ ⓓ ⓔ	1. ⓐ ⓑ ⓒ ⓓ ⓔ
2. ⓕ ⓖ ⓗ ⓙ ⓚ	2. ⓕ ⓖ ⓗ ⓙ ⓚ	2. ⓕ ⓖ ⓗ ⓙ ⓚ
3. ⓐ ⓑ ⓒ ⓓ ⓔ	3. ⓐ ⓑ ⓒ ⓓ ⓔ	3. ⓐ ⓑ ⓒ ⓓ ⓔ
4. ⓕ ⓖ ⓗ ⓙ ⓚ	4. ⓕ ⓖ ⓗ ⓙ ⓚ	4. ⓕ ⓖ ⓗ ⓙ ⓚ
5. ⓐ ⓑ ⓒ ⓓ ⓔ	5. ⓐ ⓑ ⓒ ⓓ ⓔ	5. ⓐ ⓑ ⓒ ⓓ ⓔ
6. ⓕ ⓖ ⓗ ⓙ ⓚ	6. ⓕ ⓖ ⓗ ⓙ ⓚ	6. ⓕ ⓖ ⓗ ⓙ ⓚ
7. ⓐ ⓑ ⓒ ⓓ ⓔ	7. ⓐ ⓑ ⓒ ⓓ ⓔ	7. ⓐ ⓑ ⓒ ⓓ ⓔ
8. ⓕ ⓖ ⓗ ⓙ ⓚ	8. ⓕ ⓖ ⓗ ⓙ ⓚ	8. ⓕ ⓖ ⓗ ⓙ ⓚ
9. ⓐ ⓑ ⓒ ⓓ ⓔ	9. ⓐ ⓑ ⓒ ⓓ ⓔ	9. ⓐ ⓑ ⓒ ⓓ ⓔ
10. ⓕ ⓖ ⓗ ⓙ ⓚ	10. ⓕ ⓖ ⓗ ⓙ ⓚ	10. ⓕ ⓖ ⓗ ⓙ ⓚ
11. ⓐ ⓑ ⓒ ⓓ ⓔ	11. ⓐ ⓑ ⓒ ⓓ ⓔ	11. ⓐ ⓑ ⓒ ⓓ ⓔ
12. ⓕ ⓖ ⓗ ⓙ ⓚ	12. ⓕ ⓖ ⓗ ⓙ ⓚ	12. ⓕ ⓖ ⓗ ⓙ ⓚ

© Carson-Dellosa CD-3754

4 × 2	4 × 3	4 × 4	5 × 1
5 × 2	5 × 3	5 × 4	5 × 5
6 × 1	6 × 2	6 × 3	6 × 4
6 × 5	6 × 6	7 × 1	7 × 2

5	25	24	14
16	20	18	7
12	15	12	36
8	10	6	30

7 × 3	7 × 4	7 × 5	7 × 6
7 × 7	8 × 1	8 × 2	8 × 3
8 × 4	8 × 5	8 × 6	8 × 7
8 × 8	9 × 1	9 × 2	9 × 3

© CD-3754

42	24	56	27
35	16	48	18
28	8	40	9
21	49	32	64

9 × 4	9 × 5	9 × 6	9 × 7
9 × 8	9 × 9	10 × 1	10 × 2
10 × 3	10 × 4	10 × 5	10 × 6
10 × 7	10 × 8	10 × 9	10 × 10

© CD-3754

63	20	60	100
54	10	50	90
45	81	40	80
36	72	30	70

$2\overline{)4}$	$2\overline{)6}$	$2\overline{)8}$	$2\overline{)10}$
$2\overline{)12}$	$2\overline{)14}$	$2\overline{)16}$	$2\overline{)18}$
$3\overline{)3}$	$3\overline{)6}$	$3\overline{)9}$	$3\overline{)12}$
$3\overline{)15}$	$3\overline{)18}$	$3\overline{)21}$	$3\overline{)24}$

5	9	4	8
4	8	3	7
3	7	2	6
2	6	1	5

3)27	4)4	4)8	4)12
4)16	4)20	4)24	4)28
4)32	4)36	5)5	5)10
5)15	5)20	5)25	5)30

3	7	2	6
2	6	1	5
1	5	9	4
9	4	8	3

5)35	5)40	5)45	6)6
6)12	6)18	6)24	6)30
6)42	6)48	6)54	7)7
7)14	7)21	7)28	7)35

© CD-3754

1	5	1	5
6	4	9	4
8	3	8	3
7	2	7	2